C0-AYI-742

Acid Rain Science and Politics in Japan

Politics, Science, and the Environment
Peter M. Haas, Sheila Jasanoff, and Gene Rochlin, editors

Peter Dauvergne, *Shadows in the Forest: Japan and the Politics of Timber in Southeast Asia*

Peter Cebon, Urs Dahinden, Huw Davies, Dieter M. Imboden, and Carlo C. Jaeger, eds., *Views from the Alps: Regional Perspectives on Climate Change*

Clark C. Gibson, Margaret A. McKean, and Elinor Ostrom, eds., *People and Forests: Communities, Institutions, and Governance*

The Social Learning Group, *Learning to Manage Global Environmental Risks.* Volume 1: *A Comparative History of Social Responses to Climate Change, Ozone Depletion, and Acid Rain.* Volume 2: *A Functional Analysis of Social Responses to Climate Change, Ozone Depletion, and Acid Rain*

Clark Miller and Paul N. Edwards, eds., *Changing the Atmosphere: Expert Knowledge and Environmental Governance*

Craig W. Thomas, *Bureaucratic Landscapes: Interagency Cooperation and the Preservation of Biodiversity*

Nives Dolšak and Elinor Ostrom, eds., *The Commons in the New Millennium: Challenges and Adaptation*

Virginia M. Walsh, *Global Institutions and Social Knowledge: Generating Research at the Scripps Institution and the Inter-American Tropical Tuna Commission*

Kenneth E. Wilkening, *Acid Rain Science and Politics in Japan: A History of Knowledge and Action toward Sustainability*

Acid Rain Science and Politics in Japan

A History of Knowledge and Action toward Sustainability

Kenneth E. Wilkening

The MIT Press
Cambridge, Massachusetts
London, England

This book was set in Sabon by SNP Best-set Typesetter Ltd., Hong Kong. Printed on recycled paper and bound in the United States of America.

Library of Congress Cataloging-in-Publication Data

Wilkening, Kenneth E.
Acid rain science and politics in Japan : a history of knowledge and action toward sustainability / Kenneth E. Wilkening.
p. cm.—(Politics, science, and the environment)
Includes bibliographical references and index.
ISBN 0-262-23235-9 (hc. : alk. paper)—ISBN 0-262-73166-5 (pbk. : alk. paper)
1. Acid rain—Government policy—Japan—History. 2. Environmental management—Japan—History. I. Title. II. Series.

QC926.57.J3W55 2004
363.738′67′0952—dc22

2003065175

10 9 8 7 6 5 4 3 2 1

To Tokiko,
who walked all the rains with me

Contents

Series Foreword ix
Acknowledgments xi
Notes on Style xiii
Abbreviations and Acronyms xv

1 **Sustainable Rain? 1**

2 **Sustainable Science, Politics, and Environment 11**
Packages of Science 13
The Role of Science in Environmental Policy 22

3 **Nature, Culture, and the Acid Deposition Problem in Japan 33**
The Acid Deposition Problem 33
Nature and Culture in Japan 38
Science in Japan 48
Japan's Acid Deposition History 54

4 **Period 1 (1868–1920): Copper Mines and Early Precipitation Chemistry 61**
First Environmental Era (1868–1945): Prewar Industrialization 61
Copper Mines 62
Early Precipitation Chemistry 76
Analysis of Period 1 78

5 **Period 2 (1920–1945): Flowering of Precipitation Chemistry 95**
Early Urban Air Pollution 95
Precipitation Chemistry Comes of Age 97
Analysis of Period 2 112

6 **Period 3 (1945–1974): Air Pollution Reigns 121**
Second Environmental Era (1945–1967): Postwar Reconstruction 121
Third Environmental Era (1967–1973): Domestic Environmental Revolution 126
Quiet Acid Deposition Research 133
Analysis of Period 3 141

7 **Period 4 (1974–1983): Moist Air Pollution 149**
Fourth Environmental Era (1973–1990): Transition 149
Moist Air Pollution 152
After Moist Air Pollution 158
Analysis of Period 4 163

8 **Period 5 (1983–1990): Ecological Acid Deposition Research 173**
Acid Rain Surveys 173
Analysis of Period 5 182

9 **Period 6 (1990–present): East Asian Transboundary Air Pollution 193**
Fifth Environmental Era (1990–present): Global Environment 194
Internationalization of Acid Deposition 197
Regime Formation and Regional Sustainability 222

10 **Sustainable Rain: Knowledge and Action 225**

Appendix: Japan's Present-Day Acid Deposition Science 237
Major Scientific Research Programs 238
LRT Problem-Framework 242

Notes 265
References 289
Index 315

Series Foreword

As our understanding of environmental threats deepens and broadens, it is increasingly clear that many environmental issues cannot be simply understood, analyzed, or acted on. The multifaceted relationships between human beings, social and political institutions, and the physical environment in which they are situated extend across disciplinary as well as geopolitical confines and cannot be analyzed or resolved in isolation.

The purpose of this series is to address the increasingly complex questions of how societies come to understand, confront, and cope with both the sources and manifestations of present and potential environmental threats. Works in the series may focus on matters political, scientific, technical, social, or economic. What they share is attention to the intertwined roles of politics, science, and technology in the recognition, framing, analysis, and management of environmentally related contemporary issues, and a manifest relevance to the increasingly difficult problems of identifying and forging environmentally sound public policy.

Peter M. Haas
Sheila Jasanoff
Gene Rochlin

Acknowledgments

Stately and tall, a book is like a tree. Without the ecological network that hidden supports it, the tree would not exist. So, too, with a book; without a network of people and organizations it would not exist. A vast array of individuals and organizations supported this work. I wish to acknowledge and thank all those who formed my network. This includes Professors Calvin DeWitt, Erhard Joeres, Neil Richardson, and Kirk McVoy of the University of Wisconsin, Madison, Emanuel Adler of the University of Toronto, and T. J. Pempel of the University of California, Berkeley.

Further, I wish to thank the hard-working researchers and staff at the National Institute for Environmental Studies (NIES) in Tsukuba, Japan for their generous support and encouragement during my two-year stay from July 1994 to July 1996. In particular, I wish to express my deep appreciation to my two hosts at NIES, Satake Kenichi and Uno Itsushi. Satake-san's afternoon "tea time" get-togethers were memorable not only for continually inspiring my investigations but also for the philosophical musings we endlessly pursued. Uno-san's meticulous dedication to his research (computer simulation of the atmosphere) was a constant reminder of the top quality science Japan has to offer to the world. Also, special thanks goes to Ueda Hiromasa of Kyoto University, who provided me with the introduction to NIES. In addition, dozens of individuals in Japan, too numerous to name, gave freely of their time and energy to provide me with materials, to be interviewed, or to guide me in one way or another through Japan's acid rain world. Several people, however, deserve to be singled out for special recognition—Tamaki Motonori of the Hyōgo Prefectural Institute of Environmental Science, Hara Hiroshi of the National Institute for Public Health, Okita Toshiichi of Obirin University, Yoshitake Takashi of the Forestry and Forest Products Research Institute, and Fujita Shinichi and Ichikawa Yoichi of the Central Research Institute of the Electric Power Industry (CRIEPI). Their enthusiasm for their work sustained mine.

The Japan Foundation provided me with a 1994–1995 Japan Foundation Fellowship that got this book started (thank you!), and the Institute for Environmental Studies (IES) at the University of Wisconsin, Madison and my present location, the University of North British Columbia (UNBC), provided the academic settings to finish the book. In addition to a stimulating academic setting, UNBC generously provided a grant to assist with the publication of this book.

Clay Morgan, Deborah Cantor-Adams, Dawn Wakefield, and the staff of the MIT Press are to be praised for making the publishing process friendly and efficient. I appreciate their efforts, as I appreciate the comments from the series editors and the anonymous reviewers. Their insights helped to greatly improve this work. In the end, however, its shortcomings remain my responsibility.

Finally, I wish to thank my parents, my brother Dean, and my sister Karen, for the "underground" support that only a family can provide, and especially my wife, Tokiko, whose deep reserve of Japanese patience assisted me in innumerable ways through the years that this book was in the making.

Notes on Style

Japanese words, except people and place names, are italicized unless they have become common English words such as sushi.

Macrons appear over long vowels (ū and ō) except for well-known terms such as Tokyo and Osaka.

In proper names of agencies, temples, and the like, the suffix—such as *-ji* for temple—is retained and followed with the English equivalent without a capital letter, e.g., Hōryū-ji temple.

Japanese personal names are given in the normal East Asian order: surname first, followed by the given name.

Abbreviations and Acronyms

EA	Environment Agency of Japan
EANET	East Asian Acid Deposition Monitoring Network
EMEP	European Monitoring and Evaluation Programme
EPA	U.S. Environmental Protection Agency
$g/m^2/yr$	grams per square meter per year
GEPRP	Global Environmental Protection Research Plan
GERF	Global Environmental Research Fund
kg/ha/yr	kilograms per hectare per year
km	kilometers
LDP	Liberal Democratic Party
LRT	long-range transport
LRTAP	UNECE's Convention on Long-Range Transboundary Air Pollution
MAFF	Ministry of Agriculture, Forestry, and Fisheries of Japan
mg/l	milligrams per liter (=1 ppm)
MHW	Ministry of Health and Welfare of Japan
MITI	Ministry of International Trade and Industry of Japan
mm	millimeter
NAPAP	U.S. National Acid Precipitation Assessment Program
NGO	nongovernmental organization
NIES	National Institute for Environmental Studies
NO_x	nitrogen oxides
O_3	ozone
OECD	Organization for Economic Cooperation and Development
ODA	overseas development aid
PM	particulate matter

ppb parts per billion
ppm parts per million
RAINS Regional Acidification, Information, and Simulation model
SO_2 sulfur dioxide
SO_4^{2-} sulfate ion
ton English ton (1 ton = 0.9 tonnes)
tonne metric tonne (1 tonne = 1.1 tons)
TgS tera-grams of sulfur (10^{12} grams or 10^6 tonnes of elemental sulfur equivalent; $S/SO_2 = 0.5$
TgN tera-grams of nitrogen (10^{12} grams or 10^6 tonnes of elemental nitrogen equivalent; $N/NO_2 = 0.304$ (assuming all NO_x is NO_2).
UN United Nations
UNECE United Nations Economic Commission for Europe
VOC volatile organic compound
$\mu g/m^3$ micrograms per cubic meter

Acid Rain Science and Politics in Japan

1

Sustainable Rain?

Rain sustains the land and people of Japan. In ancient Japan, rain in its myriad forms veiled the pure air—drizzles fell from the "plum rains" of early summer, tropical-like high humidities stifled midsummer afternoons, thunder and lightening storms hit on rare occasions in late summer, typhoons drenched the countryside in autumn, snow storms buried the Sea of Japan coast in winter, coastal and mountain fogs shrouded harbors and villages, hail and sleet shocked unwary travelers, sea sprays from the surrounding ocean whisked past children at play on the beaches, mists hovered above innumerable rice paddies, and blue-tinted vapors evapo-transpirated from luxuriant coast-to-coast forests. Rain, or precipitation in its multiple forms, dominated the ancient island landscape of Japan.

Rain produced an equally moisture-laden culture. Japan is a culture of rain like the Native American Hopi culture is a culture of sun or the Arctic Inuit culture is a culture of snow. There are hundreds of words for rain in the Japanese language.[1] Japan's wet rice agriculture is fundamentally dependent on abundant rainfall. Elements of clothing (e.g., *geta* sandals), housing (e.g., tile roofs), and gardens (e.g., the ubiquitous pond) reflect the prevalence of rain. In Japan the umbrella was elevated to a work of art. Japanese poetry and literature breathe rain, and Japanese art forms exude mist. A wandering poet-monk, Santōka, in a Zen-inspired *haiku* poem stated of himself what could be said of the people: *karada nagedashite shigururu yama* (flinging body, into drenched mountains). Black ink *sumi-e* paintings capture glimpses of distant figures in misty valleys. In *sumi-e*, as in many traditional art forms, colors are applied as if seen through a light rain. When used at all, they are muted, pastels being the pinnacle of brilliance. During the tea ceremony gazing quietly upon boiling water and its wafts of steam as it changes media—tea kettle to bamboo ladle to tea bowl—is to shed the cares of the outside world. In the tea ceremony, as in many traditional art forms, an ambience of intermittent rainfall seems to suffuse ritual. Japan's ancient culture can be likened to a watercolor painting whose wash was laid on with a broad brush dipped in sky-blue air and whose endless detail was stroked in with smaller, finer brushes dipped in the boundless forms of moisture (typhoon downpours, rainy season drizzles, hot spring steams, summer humidity, winter wet "peony" snowflakes).

Japanese culture was born in a land of pure air and clean rain, but the price the people have paid for a modern, industrialized, urbanized world is to lose their pristine air and natural rain. The air and rain are now polluted. In this book we will witness the historical alteration and degradation of Japan's air and rain as wrought by one class of pollutants—acidic pollutants. We will study the history and present state of modern Japan's "acid rain" problems. (Even though "acid rain" is the popular designation for the problem, the term "acid deposition" is most often used in this book. The reasons will be made clear in chapter 3.) Specifically, we will investigate the science and politics of acid rain problems from the Meiji Restoration of 1868 to the present, ending around the year 2000.

Why devote a whole book to analyzing the seemingly arcane history of acid deposition problems in Japan? There are a host of reasons.

- Japan has long experienced acid deposition problems, longer than any other country in East Asia. Acid deposition is, first of all, a millennial old natural "problem" in Japan. The existence of numerous active volcanoes since the birth of the islands creates a slight but perennial acid deposition problem. No other East Asian country, and few nations in the world, possess a greater density of active volcanoes or a greater number of acidic environmental niches due to plate tectonic activity than Japan.
- Acid deposition as a human-induced problem is over 100 years old in Japan. The first instances of the problem resulted from copper smelting around the turn of the century. At this time the first efforts at acid deposition-related scientific research and policy measures occurred. No

other East Asian country earlier attempted research or policymaking on the issue.

• Acid deposition as a postwar environmental problem was first seriously studied in Japan in the mid-1970s. No other East Asian country tackled the postwar acid deposition problem earlier than Japan.

• Japan discovered the existence of long-range transport of air pollutants in East Asia in the mid-1980s, and in the process revealed acid deposition to be a transboundary problem in the region. Japan also discovered it was a "victim" of this transboundary air pollution (i.e., it imported more acidic pollutants than it exported to surrounding countries).

• East Asia now has the dubious honor of being the world's third regional-scale acid deposition hot spot after Europe and North America. Transboundary air pollution (of which cross-boundary movement of acidic pollutants is the most prominent aspect at present) is a potentially explosive international environmental issue in East Asia. Japan is leading diplomatic efforts to defuse the issue. In the process, Japan is breaking new ground and setting scientific and political precedents in regional environmental cooperation.

• Japan is a world leader in environmental science and policymaking. It is the first Asian nation to achieve such status. What it learns in East Asia on the acid deposition issue will be translated to the world stage.

As the most scientifically advanced, economically powerful, and democratically sophisticated nation in East Asia, Japan is in a natural position of scientific, technological, economic, and political leadership on environmental issues. This guarantees that Japan's science and policymaking will have a major impact, and in many ways define, the approach to acid deposition and other environmental problems in East Asia. Understanding Japan's influence on environmental problem-solving in East Asia, and the world, provides sufficient incentive to investigate the historical roots of an environmental problem like acid deposition.

Acid deposition has a long history as an international environmental issue—over thirty-five years in Europe, over thirty years in North America, and now over fifteen years in East Asia. It was the first transboundary air pollution problem not related to nuclear weapons testing around which an international convention was signed (in 1979).[2]

Acid deposition has an even longer history as a local and national problem—over 100 years in Europe, North America, and East Asia. The long and multidimensional history of the problem makes it ripe for comparative analysis and for inquiry into long-term patterns related to environmental science and politics. Although the study contained in this book does not pursue comparative analysis (it is a single-site "archeology of knowledge"), it does seek to understand general patterns related to environmental science and politics that are not necessarily country specific.

Despite the focus on a single nation, the study is actually *international* in scope. One of the surprises revealed by analyzing Japan's acid deposition history is the strength of international influences from the very first manifestation of the problem around the turn of the century. International forces have been highly instrumental in creating, understanding, and solving the various historical manifestations of Japan's acid deposition problems. In particular, ideas drawn from Europe and North America have been especially prominent. Even though I do not attempt a comparative analysis of Japan's acid deposition history with those of Europe and North America, I do situate Japan's history within the context of larger international developments on the issue.[3] In particular, I highlight the role of Europe and North America as source regions for ideas imported by Japan, and ask questions such as: How and why were acid deposition ideas imported from Europe and North America to Japan? How were they transformed (or "localized" to Japan) upon arrival? And how have they shaped Japan's acid deposition science and policy?

Besides the fact that Japan has a long and fascinating acid deposition history that informs its present environmental science and policy, a second reason for analyzing the history is to illuminate patterns of interaction between environmental science and politics in general. Science dominates Japan's acid deposition story from beginning to present. Until the 1970s it was essentially one of scientists straining to define an object of policymaking, and since the 1970s it has been primarily one of scientists and bureaucrats shaping "science policy" (e.g., decisions to fund research). This relative simplicity makes it a perfect candidate for analyzing the science component of the policymaking process. I extracted from Japan's acid deposition history a generic set of concepts and relationships, spelled out in chapter 2, that I believe are broadly applicable to theoretical understanding of the interaction between science and politics on environmental issues wherever it occurs.

To my knowledge this book is the first to trace in detail both the science and politics of an environmental problem in Japan from its origin in the country's opening to the West in the late 1800s to the present. It is certainly the first related to the acid deposition problem. There is a vast literature on the science and politics of the acid deposition problem in the West. Very little has been written on its history in Asia. Thus, this work will contribute to the emerging field of Asian environmental history. I did not set out to write a book on environmental history, though. Indeed, I am not a historian. Although I have long had an interest in history, I am trained in environmental science and international environmental policy analysis. I began studying the science and politics of the transboundary air pollution issue in East Asia in the early 1990s, and what started out as simple curiosity as to the roots of Japan's present-day international environmental policymaking on the issue, turned, much to my amazement, into a prolonged investigation that revealed a history far longer, deeper, and richer than I ever imagined. I find it an intriguing history. A sense of dedicated puzzle-solving, humble integrity, energetic persistence, and missed opportunities pervades the story. I have tried to convey to the reader some of the human drama behind the facts, figures, and events.

"Simple curiosity" is perhaps too mild an expression for what motivated my plunge into Japan's acid deposition history. I was studying the role of science in Japan's leadership position on the transboundary air pollution issue in East Asia (initially framed as a regional "acid rain" problem), and I kept running up against a feeling that the sum total of *contemporary* factors used to explain its leadership position (such as Japan's victim status, its economic wealth, its desire to sell its environmental technology, its desire to enhance its environmental reputation, etc.) was not sufficient to comprehend its dogged and proactive pursuit of international acid deposition diplomacy. This led me to the history of the issue. What I discovered was that there are also significant *historical* factors that help explain Japan's activism. In particular, a strong "historical scientific momentum" related to the problem fortuitously placed Japan in a position to scientifically recognize and politically act upon its international dimension in the region. I do not posit that such historical scientific momentum is a prerequisite for active international environmental diplomacy, but I do argue that in Japan's case this, combined with its "culture of rain," imparts a special aura and depth of commitment to the issue that is not necessarily encountered in its other environmental problem-areas.

This book is primarily directed at scholars in the fields of environmental science (especially scientists studying acid deposition) and political science (especially researchers analyzing the role of science in environmental policymaking). However, it also targets specialists in history (especially historians of Asian science) and Japanese culture (especially students of Japan's relationship to nature). My hope is that all these academic communities will find something of value to inform their respective disciplines. The reader should be aware, though, that the book has an unavoidable bias toward the sciences. As mentioned before, analysis of Japan's acid deposition history is highly conducive to investigating the workings of science at the science-policy interface. Thus, I have chosen to lean heavily on the science side of the equation. I dig deep into the science to be able to illuminate questions of how, when, where, and why environmental scientific knowledge worked its way into the policymaking world. Digging deep means I provide a high level of detail about Japanese acid deposition science. The result may be that those not well versed in environmental science will find some portions of the book difficult reading. If so, I encourage you to persevere or to skim. I have tried to structure the text so that it is easy to navigate around highly technical sections.

Despite the specialized nature of the book, there is a larger, nonspecialist audience to whom the book is also addressed—those who desire greater insight into the modern struggle to attain *sustainability* (a definition of sustainability is provided in the next chapter). In the course of my inquiry, besides unearthing a wealth of information related to Japan's acid deposition problems, certain underlying patterns emerged that, I believe, are relevant to grappling with one of the paramount problems of modern civilization; namely, how to create ecologically and socially "sustainable societies." Thus, besides the more limited purposes already stated, this history serves another, grander purpose—to chronologically document in one corner of the world in one issue-area the human struggle to walk the path of sustainability, and to seek in this effort lessons that might be applied to the global challenge of creating a "sustainable world."

During the Edo period (1600–1868), Japan was for all practical purposes a sustainable society.[4] It was sustainable in the sense that it was self-sufficient in human and natural resources. Also, its population had leveled off by the middle of the period. There was only a trickle of people, goods, and ideas into the country, especially after a policy of national isolation (*sakoku*) was invoked in 1639. The Meiji Restoration of 1868 marked not only the end of the Edo period itself, but also the end of Japan as a self-sufficient and sustainable society. In sharp contrast to the Edo period, the ensuing Meiji period (1868–1912) witnessed a flood of foreign people, goods, and ideas. One small stream of foreign people (e.g., mining engineers), goods (e.g., mining technologies), and ideas (e.g., principles of capitalism) contributed to "acid deposition unsustainability." In other words, they contributed to the emission into the atmosphere of large quantities of acidic substances that in certain locales created socially and ecologically unacceptable and unsustainable conditions. However, another small stream of foreign people, goods, and ideas (e.g., German chemists, precipitation chemistry monitoring equipment, and precipitation chemistry techniques) created the foundation for systematic understanding of and political action on problems such as acid deposition. Since 1868 Japan has experienced ups and downs on its path toward sustainability. Certain by-products of industrialization periodically knocked it off a sustainable course and countervailing elements in environmental science and politics often tried to get it back on course. We will follow this dynamic relative to acidic pollutants in the atmosphere.

Japan has experienced three pollution-related sustainability crises since the end of the Edo period.[5] The first occurred around the turn of the century and was related to copper mining. The second occurred after World War II and was related to massive industrial pollution. And the third is the present-day, global crisis being experience by Japan and the rest of the world in which pollutants are exchanged between all nations and transported to the furthest reaches of the Earth. Since acid deposition played or plays a role in each of these pollution-related sustainability crises, investigation of acid deposition science and politics provides a window onto Japan's struggle toward ecological and social sustainability and clues to the attainment of sustainability in Japan and the rest of the world.

Japan's acid deposition history can be used to help answer questions about the origins of scientific knowledge, the influence of scientific knowledge on politics, and (un)sustainability related to environmental problems because our starting condition is a country in a sustainable state without an acid deposition problem and without a scientific tradition. Thus, Japan began more or less as a "blank slate" in relation to acid deposition science, politics, and sustainability in 1868. I use Japan's long acid deposition history to identify some of the essential aspects of science and the science-policy interface whose presence or absence helps explain political action and inaction in relation to sustainability. The historical portrait of the science-policy-sustainability triad that emerges is one of great organic complexity.

The book is composed of the following chapters.

Chapter 2 introduces the general set of concepts used to analyze the science-politics nexus. These concepts are employed in the remainder of the book to track and explain the relationship between science and policy related to the acid deposition problem in Japan.

Chapter 3 discusses nature, culture, and the acid deposition problem in Japan. It begins with a brief introduction to the acid deposition problem in general. It continues with an overview of elements of Japan's natural environment and culture that are relevant to its acid deposition problems. This is followed by a quick sketch of the history of science in Japan, which in turn serves as a preamble for describing in the final section the environmental and acid deposition chronologies used to organize analysis of Japan's acid deposition history. The swath of history between 1868 and the present (circa 2000) is divided into five environmental eras and six acid deposition periods.

Chapters 4–9 discuss in detail each of the six acid deposition periods.

Chapter 10 synthesizes and summarizes what was learned in the process of analyzing Japan's acid deposition history, and draws lessons that might be applied to the challenge of creating sustainable societies in Japan, Asia, and the rest of the world.

An appendix describes the present state of acid deposition science in Japan.

2

Sustainable Science, Politics, and Environment

From the dawn of human evolution two interrelated activities enhanced the survival of human groups—systematic observation and mental mapping of the natural world surrounding the group (primitive environmental science), and the effective ordering of priorities vital to the group in relation to the natural surroundings and authoritative execution and enforcement of actions based on those priorities (primitive environmental politics). These activities fulfill two essential needs: (1) the need to understand a vastly complex natural world, and (2) the need to manage a vastly complex social world in relation to the natural world. Phenomenal changes have taken place in the human relationship to the natural world since the emergence of our species some three million years ago, but the needs remain unchanged. The goal of (primitive and modern) environmental science is still to understand a vastly complex natural world, and the goal of (primitive and modern) environmental politics is still to manage a vastly complex social world in relation to the natural world. Today, environmental science and politics are confronted with a monumental challenge—harmonizing the complexity of the natural and social worlds so as to develop sustainable societies.

The concept of sustainability gained prominence in the 1980s, and speaks to the long-term survival of the human race. It enshrines the idea of intergenerational equity. While myriad definitions of sustainability have been put forth, reflecting the inherent ambiguity of the term, the most famous is that of the World Commission on Environment and Development (WCED): "[sustainability, or sustainable development] meets the needs of the present without compromising the ability of future generations to meet their own needs" (WCED 1987, 43).[1] My own definition parallels and elaborates on this definition, as follows.

Sustainability is the goal-oriented process of realizing integrated human/ecosystem well-being without compromising the ability of future generations and ecosystems to maintain their well-being.

Operationalizing this definition is exceedingly difficult. It involves devising new institutions, policies, ways of life, and modes of thinking based in common sense, inherited wisdom, and modern science. The concept of sustainability imparts to modern environmental science and politics a clear-cut objective—attainment of long-term, large-scale sustainability. The goal of environmental science is to assess states of unsustainability and define sustainability criteria. The goal of environmental politics is to set and implement policies to achieve sustainability. The challenge of global sustainability makes it imperative that environmental science and environmental politics maintain a positive complementarity—good environmental policy requires sound science, and sound science requires astute political support. Tightly integrating the results of scientific research with political decision making in service to sustainability is essential to the proper functioning of both science and politics. As Gro Harlem Brundtland, the head of the WCED, states: "As caretakers of our common future, we have the responsibility to seek scientifically sound policies, nationally as well as internationally. If the long-term viability of humanity is to be ensured, we have no other choice" (Brundtland 1997, 457).

We can distinguish between ecological and social sustainability. Ecological sustainability (ecosystem well-being) is "the existence of the ecological conditions necessary to support human life at a specified level of well-being through future generations" (Lélé 1991, 609). There is no clear definition of social sustainability (human well-being). However, it loosely refers to providing basic needs for food, water, clothing, shelter, health care, and education; to offering opportunity to attain a satisfying quality of life; to maintaining the viability of cultures and local communities; and to supporting social, economic, and political infrastructure for delivering goods and services (Lélé 1991, 615). This book focuses primarily but not exclusively on ecological sustainability. One of the tasks of (natural) scientists is to spell out the biogeochemical "laws" underpinning social and ecological sustainability. In turn, the constraints implied by these natural laws are fed into the political process so that value judgments can be made on what is to be sustained, for whom, and for how long.

Historically, how have environmental science and politics interacted relative to the goal of sustainability? Can we glean from their historical relationship "sustainability lessons" for the future? In this book, I use Japan's acid deposition history as a database for investigating the science-politics interface in relation to sustainability. There were two interrelated dimensions to my investigation.

The first was devising a generalized set of concepts situated at the science-policy interface for analyzing and understanding the interaction between the scientific and political worlds vis-à-vis sustainability. This set of concepts forms the theoretical core of the book and is explained in the rest of this chapter. Many of the concepts are informed by or adapted from the science studies literature.[2] The second dimension was using my conceptual toolkit to uncover large-scale, historical patterns relative to the science-politics-sustainability triad. The bulk of the book focuses on bringing these patterns to light in relation to Japan's acid deposition problems, specifically to answering three central questions: (1) Throughout its modern history, how and why has Japan gone about acquiring scientific knowledge related to acid deposition phenomena? (2) How, and under what circumstances, have scientists and scientific knowledge influenced public policymaking in Japan on the acid deposition issue? (3) If influential, did this steer Japan toward acid deposition sustainability, why or why not?

Packages of Science

Science is both a body of knowledge and a method for obtaining it. And, science is scientists, the practitioners of the methods that generate the knowledge. Thus, science can be viewed as "knowledge" and as "people and practice." This distinction is crucial to our understanding of the science-politics interface. Science (science-as-knowledge and science-as-practice) is often portrayed as homogeneously constituted, ahistorical, and linearly cumulative in its knowledge base. Contemporary analysis of the meaning and practice of science has exposed the fallacy of this vision. Philosophers, historians, sociologists, political scientists, anthropologists, and other scholars point out that science has decidedly nonuniform, nonuniversal, and even quirky characteristics in terms of both its knowledge and practice. Science proves to be patchy. Many scholars argue, for example, that scientific knowledge and, by association, scientists come bundled in historically changing units. Fragments of scientific knowledge have little meaning outside the context of their accompanying unit, and groups of scientists derive their coherence of practice from their affiliated cohort. I join this tradition and argue in this book that science in relation to environmental politics comes in historically changing units, one associated with scientists and the other with scientific knowledge.

Perhaps the most famous historically changing, scientific knowledge-related unit is Thomas Kuhn's (1962) "paradigm." A paradigm is a scientific model or pattern (that may consist of scientific laws, principles, assumptions, applications of laws, and experiments or observations of the behavior of nature) that a particular scientific community acknowledges for a historically defined period of time as supplying the foundation of its further practice. According to Kuhn, the process of scientific discovery is divided into periods of normal science, which operate under a given paradigm, and revolutionary science during which a major transition occurs to a new paradigm that in turn becomes the standard for the next period of normal science. Another scientific unit developed in counterpoint to Kuhn's paradigm is Imre Lakatos's (1970) scientific "research program." A research program is a series of connected theories that develops internally. It contains a "hard core" of fundamental ideas about a segment of reality and a "protective belt" of theories by which the hard core is interpreted. The protective belt bears the brunt of tests and is adjusted or even replaced over time. Another scientific unit, applicable also to the world beyond science, is Michel Foucault's (1970, 1972) "episteme." An episteme is a global thought-form configuration, a universal system of reference, submerged deep under the ocean of all theoretical knowledge of an age that confers upon that knowledge its specific historical character. In one of his more famous statements, Foucault wrote: "In any given culture and at any given moment, there is always only one episteme that defines the conditions of possibility of all knowledge, whether expressed in a theory or silently invested in a practice" (Foucault 1970, 168).

Problem-Framework

In the course of my research on the history of acid deposition science in Japan, I detected an historically changing scientific unit kindred to paradigm, research program, and episteme; a unit that is relevant to the politics of an environmental problem. It is not as grandiose as paradigm, research program, or episteme. Nor does it necessarily have to do with the discovery of new scientific knowledge, which is generally their preoccupation. While it involves scientific discovery, it also relates to the more prosaic processes of representation, transmission, and transformation of known scientific knowledge as it moves across boundaries (national, social, and cultural).

According to what I saw in the history of acid deposition science in Japan, there are historically changing "scientific knowledge packages," or "problem-frameworks" as I call them, that scientists construct (or attempt to construct) to frame analysis and understanding of an environmental problem. Problem-frameworks weave threads of newly discovered scientific knowledge and threads of old knowledge imported from elsewhere to represent the problem at hand. Problem-frameworks may inform the problem-solving process, and they may or may not contain statements related to sustainability. In modified form, these frameworks (and their sustainability statements) can be communicated to, and influence, political decision making. The process by which this takes place is discussed later. In the case of Japan, the history of acid deposition science exhibits recurring periods during which different problem-frameworks were constructed. During some periods the frameworks influenced the policy world; during others they never made it that far. My task was to figure out how and why problem-frameworks were constructed in some periods and not in others; and if constructed, how and why they influenced politics and whether and in what form they addressed sustainability.

I define a scientific framework related to an environmental problem (or *environmental problem-framework*, for short) as:[3]

a more or less coherent, codified, and integrated set of methodologies, knowledge derived from the methodologies, and orientations and approaches to the generation and application of the knowledge, that relate to a specific environmental problem area, and that serve as a template for researching and understanding the problem and for action to address the problem area during a given time period and at a given geographical location.

Before continuing, a word needs to be said about the difference between a "problem" and a "problem-framework." A problem refers to a segment of underlying reality, and a problem-framework to a socially constructed image of that reality. A problem is never fully knowable. Thus, there may be multiple problem-frameworks of a single problem. This may even precipitate a "battle of problem-frameworks" if their representations of underlying reality significantly conflict.

An environmental problem-framework contains five key elements, as follows.

1. It relates to a specific environmental problem area, such as acid deposition, that requires at least in part interpretation by scientists and technical specialists. Thus, it defines what I call a "*focus-problem.*" The problem-framework provides a template for understanding and taking action on the focus-problem (the action element will be discussed further below in relation to a problem-framework's "solution path").

2. A problem-framework contains a set of expert methodologies for generating knowledge related to the focus-problem. In the process of transmission to other scientists, methods are transferred along with a knowledge base. The methodologies are critical to applying knowledge to new locations.

3. A problem-framework contains a specialized knowledge base or "knowledge-world."[4] Application of expert methodologies yields a more or less organized pool of knowledge that is the core of a problem-framework. The knowledge base is made up of observational data, uncontested cause-and-effect statements, well-documented theories, *and*, conjectural arguments, emerging hypotheses, unresolved problems, conflicting data sets, etc. While at once a coordinated and structured "objective whole" built of standardized elements, it also has attached to it a collection of

uncertainties, gaps in analysis, disputed areas, and so forth. In short, it consists of a consensus core and a contested periphery. The contested periphery is the principle zone of controversy.

4. A problem-framework contains an overall orientation or approach for viewing the given focus-problem. This may be determined by larger forces outside the realm of the problem area itself and thus constitutes the context for the framework. For instance, Japan's acid deposition problem-framework shifted from a human health to an ecological orientation in the 1970s.

5. A problem-framework is applicable to a specific time and place. In other words, it is an historically and geographically bound entity. It delineates the concrete manifestation of natural phenomena at a specified time and location. Thus, it changes historically and geographically.

My problem-framework concept was in part informed by Joan Fujimura's (1992, 1996) "standard package." Fujimura seeks to understand how collective scientific work is managed across social worlds (e.g., across multiple disciplines, research institutions, funding agencies) to support the production of knowledge.[5] Her central concept for making sense of how multiple social worlds come to support a particular field of research is a standard (scientific) package that she defines as a scientific theory and a standardized set of methodologies that are adopted by members of the multiple social worlds to construct a temporarily stable definition of the problem area and a thriving line of research on it (Fujimura 1992, 169). Thus, a standard package is another scientific knowledge-based unit akin to paradigm, research program, and episteme.

Fujimura initially developed her concept to understand "why and how the molecular biological bandwagon in cancer research developed" (Fujimura 1992, 176). She argues that "[a standardized package consisting of] a scientific theory and a standardized set of technologies . . . succeeded in enrolling many members of multiple social worlds in constructing a new and at least temporarily stable definition of cancer" (Fujimura 1992, 176–177). To develop a standardized package, though, something is needed to facilitate communication between the multiple worlds that deal with the particular line of research in question. That something is a "boundary object," a concept she borrows from Star and Griesemer (1989). I also use the boundary object concept; however, I have relabeled it "bridging object." A definition is given later. In essence, standard packages and boundary objects serve to organize and manage relations between all the social worlds (scientific and otherwise) involved in the generation of scientific knowledge on a given topic.

Many of the concerns that inform Fujimura's work are similar to mine—the generation, transmission, and transformation of scientific knowledge; the link between science-as-knowledge and science-as-practice; and the linking of multiple social worlds in the pursuit and use of scientific knowledge. Although we share many concerns, there are differences. For one, she focuses on the nitty-gritty production of scientific knowledge and examines "life in the laboratory." She holds a microscope to the process of knowledge production, joining a long line of other scholars (see, for instance, Latour and Woolgar [1986] and Hacking [1983]). I am one or two steps removed from this basement-level analysis. I am ultimately interested in how produced knowledge filters into, and is utilized by, the political world. Fujimura's political focus is restricted. She examines how politics shapes the mechanics of knowledge generation in terms of funding, institutional support, and so on, but does not address the question of how generated knowledge influences the larger world of actors who participate in political decision-making processes. Despite the similarities between a problem-framework and standard package, I chose not to use Fujimura's terminology because of its singular focus on the production of scientific knowledge. While I agree with her assessment that some sort of standardized package or framework serves as an essential organizing device for generating support for knowledge production activities (i.e., that scientific knowledge is co-constructed), her package seems not to make sufficient distinction between the work of credentialled scientists and work of the support apparatus.[6]

Universal and Local Science

Multiple environmental problem-frameworks are often constructed on the same focus-problem. Why? One reason derives from the simple fact that (1) environmental problems are typically heterogeneous (i.e., they are not globally uniform in character) and (2) political jurisdictions are also heterogeneous (i.e., they are also not globally uniform in character). Thus, environmental problems and political systems exhibit different characters in different localities. Geographically specific problem-frameworks must be constructed to "localize" scientific knowledge so that it is relevant to a given political jurisdiction.

I distinguish two categories of environmentally related scientific knowledge that are critical to the meaning, use, and creation of problem-frameworks: "universal knowledge" and "local knowledge." *Universal* knowledge is space-time independent knowledge; in other words, knowledge that is universally applicable to any place and time on Earth. *Local* knowledge is space/time-specific knowledge; knowledge that is applicable only to a specific location and/or time period. (Caution needs to be exercised in taking the universal attribute too literally. Rather than unequivocally universal, something more along the lines of "widely applicable" is implied.)

For instance, the underlying physics of synoptic wind flow trajectories that carry air pollutants is universally applicable anywhere on Earth; however, to know East Asian synoptic wind flows, that knowledge must be applied to a specific location, in this case East Asia. The underlying physiology of root uptake of nutrients by trees is universally applicable anywhere on Earth; however, to know the uptake patterns of the Japanese cedar, that knowledge must be applied to a specific location, the Japanese islands. Though the distinction between "universal" and "space/time-specific" knowledge is simple enough, it has tremendous political implications when it comes to dealing with a given environmental issue. It is invariably space/time-specific knowledge that is the locus of political debate. For example, it is the specific emission sources, the specific monsoon climatic patterns, and the specific impacts on native plants in East Asia that are the prime objects of acid deposition scientific research and political debate in the region, not the universal patterns. Although the universal knowledge provides the foundation for specific knowledge, universal knowledge is not usually at the heart of policy deliberations.

Once knowledge is created at the frontiers of science, it must be transmitted to other scientists. Scientific knowledge is not instantaneously known to or useable by scientists around the world. It must somehow be communicated and absorbed in new locations. Universal facts, concepts, theories, and methods become "scientific bridging objects" (defined later) that can be conveyed across boundaries and transformed at new locations for the purpose of constructing localized problem-frameworks. Another way to phrase this process is to say that localized problem-frameworks are constructed from the universal elements contained in a standardized or *standard problem-framework*. A standard problem-framework is a global scientific template for scientific analysis and understanding of the environmental problem in question. It consists of those elements of scientific research conducted at various times and locations that prove to hold true independent of time and location, and therefore serves as a mechanism of scientist-to-scientist communication. In short, it is a collection of scientific bridging objects, or what I call *science-to-science bridging objects*.

In conclusion, as far as the relationship between science and policy is concerned, one of the key steps in making scientific knowledge politically relevant is rendering it applicable to the relevant political jurisdiction. This is a nontrivial and seldom recognized task. A localized environmental problem-framework is a representation of scientific understanding of the environmental problem for the given political jurisdiction (subnational, national, or international) that makes policy on the problem.

Epistemic Shift

Problem-frameworks are dynamic entities; new elements are continually being added and subtracted. However, every once in a while a major shift in the knowledge structure can occur. In my analysis of Japan's acid deposition history, I found that problem-frameworks on occasion underwent *epistemic shifts*. In other words, major changes in the methods, knowledge, and orientation of the framework periodically transpired. Anthropogenic acidification of the atmosphere in Japan is the generic problem area of our history. We will see that between 1868 and the present five epistemic shifts occurred in the problem-framework associated with this problem area. (Indeed, I organized Japan's acid deposition history on the basis of these epistemic shifts.) Generally, the shifts resulted in frameworks that progressively incorporated more phenomena on a larger scale and with a longer time horizon than previous ones, but not always. An epistemic shift may or may not result in greater political relevance of the framework.

Expert Community

A companion scientific package to problem-framework is that of an "expert community." Problem-frameworks are crafted by expert communities, and expert communities become invested in a problem-framework. My definition of expert community is a modified version of the "epistemic community" concept commonly found in the social science literature.[7] I define a problem-specific expert community as:

a collective of experts who share common notions of what constitutes valid knowledge, and who engage in a common "problem project."

An "expert" is someone credentialled by society as such; having, for instance, received a certain diploma or certification. An expert community is a community simply by the fact of sharing: (1) notions of what constitutes valid or "certain" knowledge, and (2) a common goal of understanding a given problem area. Members subscribe to internally defined sets of criteria as to what constitutes legitimate knowledge within their fields, and, consciously produce practical knowledge and interpretations based on that knowledge pertinent to the problem domain. Thus, an expert community is more than a random and arbitrary collection of experts who happen to address a given problem. It is a collective of experts who have a purpose—to create a scientific template for understanding and action. One of the tasks of an expert community is to invent bridging objects that foster communication between scientists and that translate technical terminology for lay audiences.

Similar to problem-framework, expert communities are historically changing units. There is also a tradition in the field of science studies to identify such entities. Examples include the above-mentioned epistemic community concept; also, there are the adherents to a Kuhnian paradigm and the participants in a Lakatosian research program. The "actor-network" of Bruno Latour (1987, 1988) can also be considered a historically changing scientific grouping.

Although independent and individual experts (i.e., those without a conscious identification with other like-minded experts) may contribute to understanding an environmental problem, the work of constructing a problem-framework is generally outside the ability of a single individual. Thus, problem-framework construction is a team effort. As a problem-framework takes shape it fosters the formation of an expert community. And likewise, formation of an expert community fosters commitment to the problem-framework. This does not imply that the expert community operates in perfect harmony, though. As mentioned before, a problem-framework is characterized by a consensus core of knowledge surrounded by a contested periphery of knowledge. Differing and even competing interpretations can emanate from the contested periphery. However, the consensus core and the fact of association with attempts to understand the problem is what gives an expert community its conscious identity and a coherence that often empowers it in the political arena if and when the focus-problem assumes political importance.

The Role of Science in Environmental Policy

In this section, the relationship between the set of concepts described previously and the environmental policy process is discussed.[8] In addition, several new concepts are defined that enhance our ability to analyze the science-policy interface.

Environmental Policy Process

Politics has been defined as the "authoritative allocation of values."[9] It is attention to what is important to a group; it is adjustment to the fact of limits to natural and social resources, implying that everything that is construed as important to a group cannot be realized, and as such, matters of importance must be prioritized; and it is outcomes where in the end some people get more and some get less of important social commodities (wealth, security, happiness, and so on). Policymaking is one set of activities within the realm of politics. Policy generally relates to the conscious statements and actions of governments. It is "authoritative aspirations, internal to a government, about outcomes" (Allison and

Figure 2.1
Environmental policy process

Halperin 1972, 46). Application of these definitions of politics and policy to environmental (and sustainability) problems yield *environmental* politics and policy.

An idealized version of the environmental policy process (at any level—local, national, or international) is illustrated in figure 2.1. Although the figure captures some elements of the environmental "policy cycle," the actual policy process is neither as neat and rational nor as linear as the figure would suggest. Therefore, figure 2.1 should NOT be interpreted as a representation of all the pieces in the process nor the inevitable chronological sequence of events in the real policy world. However, it captures certain key components and a theoretical chronology that are useful in visualizing the science-policy nexus, and will be employed to help organize our understanding of the historical development of Japan's acid deposition science and policy.

The idealized policy process begins with the discovery or recognition of the existence of an environmental problem that potentially demands political action. Scientists are often the discoverers of the problem, and may or may not be able to immediately provide a problem-framework for understanding it. Simple awareness of a problem, though, is rarely sufficient to ensure political attention. If and when sufficient recognition among a variety of political actors develops, the problem can become a politically salient "issue." If adequate momentum builds behind the issue it may eventually be placed (set) on the agenda of an authoritative political body (a city council, a state legislature, a national diet, or an international organization) for discussion.

Discussion can lead to bargaining and negotiation over various policy options for addressing the problem. Eventually the body selects one policy option (a resolution is passed, a law is enacted, an international treaty is signed). Once selected, the next step is to operationalize the policy. In other words, mechanisms to maintain the policy and ensure compliance to its dictates are implemented. Once operationalized, the policy hopefully solves or at least mitigates the original environmental problem (and aids in achieving sustainability). Scientists and scientific knowledge can play a role in any or all stages of policymaking. The policy process outlined in figure 2.1 is used at the end of each chapter describing one of Japan's acid deposition periods as a device for illustrating the role and influence of science on policymaking during the period.

Two types of policymaking are considered in this book—domestic and foreign. In relation to acid deposition, domestic policymaking focuses on the domestically created manifestation of the problem, and foreign policymaking focuses on the transboundary dimension of the problem. For most of its history, Japan's acid deposition policymaking, when it occurred, was domestic decision making. Foreign policymaking did not appear until around 1990. During the 1990s Japan assumed a leadership role in pursuing international cooperation, or *regime formation*, on the acid deposition issue.

"Regime" is a technical term used in the field of international relations and will be employed later in the book. It principally, but not exclusively, relates to the process of international cooperation. Definitions, however, vary. They range from "soft" to "hard." The most famous and commonly cited (soft) definition is: "[regimes are] sets of implicit or explicit principles, norms, rules, and decision-making procedures around which actors' expectations converge in a given area of international relations" (Krasner 1983, 2). This definition engendered vigorous debate over how a regime (or international cooperation) is to be identified. Some scholars argue that observed behavior, not just rules, determines the existence of cooperation. One behavioral or hard definition (Efinger, Mayer, and Schwarzer 1993, 255) is summarized as follows: a regime exists if two conditions are met—(1) actors (principally, nation-states) in an issue-area agree on a set of principles, norms, rules, and procedures to govern their behavior with respect to the issue-area in question, and (2) the rules thereby created prove effective in the sense they are by and large complied with by the actors. Other scholars walk a fine line between hard and soft regime definitions. Robert Keohane, for instance, defines regimes as "institutions with explicit rules, agreed upon by governments, that pertain to particular sets of issues in international relations" (Keohane 1989, 4), in which the governments "recognize these agreements as having continuing validity" (Keohane 1993, 28). The regime terminology will not be used until chapter 9 when Japan's pursuit of regime formation is described in terms of its pursuit of an international treaty on transboundary air pollution in the East Asia.

Bridging Object

Expert communities and problem-frameworks serve policy needs because they address a problem that is actually, or potentially, politically relevant. The existence of an expert community and problem-framework, however, does not determine policy, negate controversy, nor does it preclude rejection of scientific advice. However, their existence does provide intellectual resources that can be called upon to help guide the environmental policy process. This brings us to the question: How exactly do expert communities and problem-frameworks influence the policy process?

To help answer this question, we return to Susan Leigh Star and James Griesemer's (1989) "boundary object" referred to previously. Boundary objects are material and nonmaterial entities that "inhabit several intersecting worlds . . . and satisfy the informational requirements of each of them. Boundary objects are objects which are both plastic enough to adapt to local needs and constraints of the several parties employing them, yet robust enough to maintain a common identity across sites.

They are weakly structured in common use, and become strongly structured in individual-site use. They have different meanings in different social worlds but their structure is common enough to more than one world to make them recognizable, a means of translation" (Star and Griesemer 1989, 393). I have adopted the boundary object concept basically as Star and Griesemer define it and use it for analyzing the communication of scientific knowledge between the scientific and political worlds; however, I relabeled it "bridging object" to emphasize the fact that it is a boundary-*crossing* object. I define a bridging object as:

a tangible or intangible object (fact, concept, method, procedure, research institution, etc.) that transects two or more social worlds (disciplines, agencies supporting research activities, groups of political actors, cultures, nations, etc.), and serves as a mechanism of communication between the worlds, and serves to structure the perceptions of members occupying the worlds, by maintaining a weakly structured common identity across the worlds and strongly structured local identity within any given world, such that it satisfies the informational requirements of each of the worlds.

A problem-framework consists of specialized knowledge and methods that, at the workaday level, are generally opaque to the untrained outsider. However, using bridging objects scientists can render their work more transparent to those who make political decisions.[10] The bridging object concept gives us leverage in understanding the variety of potential outcomes at the science-policy interface, such as harmony between the scientific and political worlds relative to knowledge production (e.g., support for research) and controversy in the political world (e.g., different actors using the same scientific knowledge for radically different political ends).

I employ two classes of bridging objects to understand the science-policy nexus: (1) those used to communicate between scientists (*science-to-science bridging objects*) and (2) those used to communicate between the scientific and political worlds (*science-to-policy bridging objects*). The first type is often a necessary prelude to the second type. Ultimately, my interest is in the science-to-policy bridging objects, since these most directly aid in understanding the political influence of science.

The main argument put forth here is that an essential mode of scientific influence on policymaking is translation of a problem-framework into lay language by expert community members using science-to-policy bridging objects. In other words, expert community members construct bridging objects that render the technical terminology, methodological minutiae, and intricate chains of reasoning of a problem-framework into a form understandable to non-experts. They provide non-expert audiences with a cognitive structure for comprehending technically complex environmental problems. These bridging objects can be broad or specific, and once invented, can be used by expert community members themselves, or other actors, to communicate between the social worlds engaged in the policy process.

Science-to-science bridging objects are the shared language of scientists, and are used, for instance, to locally adapt scientific knowledge imported from outside locations and to produce comparable and compatible data. Science-to-policy bridging objects are the "shared language" (though not necessarily shared interpretation) of political actors, and are used, for instance, as referents for political actors to make utility calculations (i.e., to calculate their interests, positions, etc.) relative to the given environmental problem. An example of both classes is the concept of pH. It is a universal scientific concept (i.e., it is space/time-independent and can be measured by scientists anywhere in the world). Even though the concept originated in Europe, as a science-to-science bridging object Japanese scientists have utilized it to produce "local" knowledge about acidic precipitation in Japan. They have measured pH values in Japan since the early part of the twentieth century. In addition, even though pH is a highly sophisticated technical concept, it can be rendered understandable to an educated lay audience. The Japanese public, for instance, has an extremely good grasp of the pH concept due to the availability of inexpensive pH meters and widespread promotion of environmental education programs and events during which it is measured. Thus, pH is also a science-to-policy bridging object; in this case influencing one important actor in Japan's acid deposition debate, the general public.

How do you tell a bridging object when you see one? Sometimes they are easy to identify, sometimes not. I used a number of techniques for identifying key bridging objects. First, personal immersion in the science of acid deposition sharpened my ability to see science-to-science bridging objects. Textbooks, for instance, were a good source for pinpointing them; once a concept has made it as far as being placed in a textbook you can be fairly confident it has a strong scientific bridging quality. Second, scientific papers often described science-to-science bridging objects, and the introduction and conclusion sections often contained condensations of research that constitute science-to-policy bridging objects. Third, interviews with scientists and political actors revealed both classes of bridging object. Fourth, final reports of scientific projects, especially their executive summaries, were a good place to find science-to-policy bridging objects. And fifth, newspaper articles proved to be an excellent repository of science-to-policy bridging objects. To positively identify a bridging object it is necessary to recognize multiple instances of the same object being used. This is sometimes tricky for science-to-policy bridging objects because different people express the same concept in different words.

The existence of science-to-policy bridging objects in and of themselves tells us little about political influence. Central to understanding political influence is classifying different types of bridging objects because different objects (representing different types and degrees of scientific knowledge) carry different political weight. Based on Japan's acid deposition history, I distinguished the following classes of environmentally related, policy-influencing science-to-policy bridging objects:

- *existence* bridging objects—concepts expressing the existence of a problem or some aspect of it without necessarily confirming its extent, intensity, cause, or impacts;
- *character* bridging objects—concepts expressing the character of some aspect of a problem;
- *extent-intensity*, *trend*, or *distribution* bridging objects—concepts describing the extent and/or intensity of the problem or an aspect of it, including trends and distributions;
- *cause-and-effect* bridging objects—concepts codifying chains of cause and effect;
- *impact* bridging objects—concepts delineating impacts of the problem on society and/or ecosystems (one type of impact bridging object is closely tied to risk assessment, called *risk assessment* bridging objects);

- *solution* bridging objects—concepts expressing all or part of the solution path (to be explained further later).

Bridging objects don't exist in a vacuum. They exist within a context that helps define their political influence. One element of context that is particularly pertinent to our investigation is culture. Cultural factors linked with bridging objects may influence policy. We will keep tabs on elements of Japanese culture, and the broader political context in Japan, within which bridging objects are situated in order to gauge their power to influence.

To understand the influence of science on the political process, ultimately we need to know what scientific knowledge various political actors use and how they interpret and deploy it. The concepts of bridging object, classes of bridging objects, and bridging object context give us tools with which to penetrate this veil.

Expert Community Activism

Another factor shaping a bridging object's political influence is expert community "activism." I distinguish three major groupings within an expert community: the policy-inactive, the policy-active, and the policy-activist. The policy-inactive remain aloof from political machinations and focus only on their knowledge generation activities. Their influence is via their ideas floating freely in the world, not their personal political involvement. The policy-active, on the other hand, participate in the policy process in some manner (for example, as members of advisory boards or negotiating teams) but do not necessarily advocate a specific value-based approach to solving a problem. The third group, the policy-activist, are those who coalesce around a given policy objective. What bonds this group is a "value-based rationale for social action."[11]

In this book, I define activism primarily in terms of the ability of experts to create, promote, and interpret science-to-policy bridging objects in the policy world. Activism is affected by, for instance, the nature and number of those in the policy-inactive, policy-active, and policy-activist subgroups, and by the number and types of activities the community engages in, such as sponsoring scientific conferences, lobbying for research funding, advising bureaucrats, appearing at hearings, holding press conferences, and so forth.

Table 2.1
Concepts for analyzing the science-policy interface

Problem-Framework:
A package of methods, knowledge, and overall orientation related to a focus-problem.
Standard vs. Localized Problem-Framework:
A standard problem-framework contains universal or space/time-*independent* elements; a localized problem-framework contains space/time-*dependent* elements, especially knowledge specifically tied to a specific geographical location and historical time period.
Epistemic Shift:
A major transformation in the methods, knowledge base, and/or orientation of a problem-framework.

Expert Community:
A credentialled expert group whose purpose is to analyze and explain the focus-problem; i.e., to create a scientific problem-framework and bridging objects that lay translate the problem-framework.
Activism:
Degree of involvement in the policy process (inactive, active, activist), and ability to create, promote, and interpret science-to-policy bridging objects.

Bridging Object:
A tangible or intangible object that transects two or more social worlds (the focus in this book is on the scientific and political worlds) and serves as a mechanism of communication between the worlds and a device for similarly structuring the perceptions of members of these worlds.
Science-to-Science Bridging Object:
A bridging object that serves as a mechanism of communication between scientists; 'universal' science (for instance, that contained in a standard problem-framework) is a collection of such bridging objects.
Science-to-Policy Bridging Object:
A bridging object that serves as a mechanism of communication between scientists and political actors engaged in the policy process.
Solution Path:
A class of science-to-policy bridging objects that provides a guide to solving or mitigating the problem; ideally, a solution path points toward SUSTAINABILITY.

Solution Path and Sustainability

A special class of science-to-policy bridging objects relate to what I call a problem-framework's *solution path*. The ultimate motive for constructing a problem-framework relative to an environmental problem is to develop courses of action to solve or mitigate the problem. Thus, besides methodologies, a collection of scientific knowledge, and an orientation for mentally orienting the scientist in use of the methods and interpretation of the knowledge, a problem-framework can also contain, if sufficiently mature, action-oriented guidelines for solving the problem. This is the solution path. A solution path does not dictate the solution; it merely circumscribes the approach. It is a low-resolution map suggesting the proper direction toward problem solution. Solving the problem will invariably involve other social, economic, political, diplomatic, and historical factors that lie outside the realm of the scientific problem-framework. For instance, the solution path to the acid deposition problem dictates that emissions into the atmosphere from major sources such as coal-fired power plants and automobiles must be lowered, but it doesn't contain a statement of what sorts of controls must be used to lower the emissions.

Sustainability enters the science-policy picture through the solution path. The solution path may contain explicit and implicit criteria for social and ecological sustainability that, in lay-translated versions, provides direction for "sustainability decisions." For instance, in the case of acid deposition a solution path may specify the amount of acidic inputs a given ecosystem can tolerate without damage. Thus, the knowledge production activities of an expert community not only can elaborate the knowledge base upon which societies make value choices, but also can circumscribe and propel solutions that direct them toward sustainability.

Table 2.1 summarizes the suite of science-related conceptual tools that were introduced in this chapter and that are used in the rest of the book to analyze Japan's acid deposition science and politics.

3

Nature, Culture, and the Acid Deposition Problem in Japan

Japan has a millennial-old, plate tectonic association with acid deposition; however, acid deposition as an anthropogenic environmental problem did not appear until Japan's opening to the West around 1850. This chapter introduces acid deposition-related nature and culture in Japan and sets the stage for analyzing Japan's acid deposition history. It begins with a brief introduction to the acid deposition problem.

Politically, the generic acid deposition problem has little meaning if it is not tied to a specific geographic setting. The next section presents the setting for this book—Japan. Japan's natural environment is the filter for localizing its acid deposition problem-framework and uniquely defines its acid deposition problem as different from anywhere else in the world. In addition, Japan's natural landscape is the wellspring from which certain aspects of Japanese culture originated that in turn shape some of its acid deposition science and politics. These aspects are also discussed.

The following section prefaces Japan's scientific encounter with acid deposition problems by briefly summarizing the historical development of science in Japan from early times to the mid-1800s, ending with the dawn of the Meiji period. The Meiji period (1868–1912), the first in Japan's modern history, is where our acid deposition story begins. The final section presents the chronological divisions used to analyze Japan's acid deposition science and politics from 1868 to the present.

The Acid Deposition Problem

Acidification of the environment as an environmental problem is any human-induced alteration of natural biogeochemical cycles that results in unnatural acidification of the natural world. Acidification in and of itself is not necessarily a problem. In the course of time many soils naturally acidify due to leaching of base cations; many lakes naturally acidify due to decomposition processes; and the soils and waters around volcanoes are often naturally acidic. However, human-induced acidification is a problem because of the rates and quantities of acidic substances artificially injected into the environment. There are a variety of ways that acidic substances can be artificially injected into the environment. They can, for instance, be released into the air, into waters, or onto soils.

The only form of the human-induced acidification addressed in this book is atmospheric acidification (i.e., the emission of acidic substances into the atmosphere, their transport through the atmosphere, and their impacts after deposition onto surfaces). However, there are other forms of human-induced acidification. For instance, changes in ground cover can induce acidification (e.g., deforestation of tropical rainforests can accelerate acidification of the soils), and injection of acids into water bodies can induce acidification (e.g., acidic effluent from manufacturing plants can acidify lakes and streams). On a global scale, atmospheric acidification (or acid deposition) is the dominant form. The first instances occurred on a very small scale ages ago as a result of activities such as forest and grassland burning, and primitive smelting, especially of copper. However, the appearance of the problem on a large scale did not occur until after the industrial revolution.[1]

Acid deposition goes by many names—acid rain, acid precipitation, acid deposition, acidifying deposition, and acidic deposition. *Acid rain* was how the problem was first designated, and is still the more popular and emotionally evocative term. Though more popular, it is not as scientifically accurate as the other terms. First of all, rain is only one form of acidified precipitation. Precipitation containing acids can occur in the form of rain, snow, sleet, hail, fog, and so forth. Second, precipitation is not the only way that acids can get from the atmosphere to the Earth's surface. Deposition can be dry (direct deposition) or wet (precipitation). Third, it is not only acids that are deposited, but also substances that are not themselves acids but that can cause acidification after deposition, i.e., acidic or acid-related substances. Thus, acidifying deposition or acidic deposition are more scientifically accurate designations. Even so, I will perpetuate a slight misnomer by consistently referring to the problem as *acid deposition*.

Acid deposition is essentially a problem of deposition of substances that result in the release of hydrogen ions (H^+ or protons). The primary source of protons is sulfuric and nitric acids that are secondary pollutants formed from primary pollutants such as sulfur oxides (SO_x) and nitrogen oxides (NO_x). But why is the release of protons a cause of adverse effects? The short answer is that protons have the power to dislodge other elements out of their place in chemical bonds. Hydrogen is the simplest element. Its atoms consist of only one proton and one electron.

When stripped of the electron the remaining hydrogen ion, or proton, is smaller than any other ion by several orders of magnitude. It is this extremely small size, combined with its electrical charge, that allows it to bond tightly to other elements. Since its bonding power is so much greater than other ions, it can displace them in molecular bonds. This displacement often results in the adverse effects associated with acid deposition. For instance, protons (hydrogen ions) can displace nutrient elements such as calcium and magnesium in soils from their retention sites on clay or mineral particles leaving them exposed to be leached out of the soil and rendered unavailable to plants for growth. As another example, protons can displace heavy metals, such as mercury or lead, from soils, lake bottom sediments, and water pipes, mobilizing them to work havoc upon whomever ingests them. While displacement by protons is not the only adverse effect of acidic substances deposited from the atmosphere, it is the dominant one.

Acid deposition is a by-product of industrialization in general, and energy production in particular. Therefore, it is a problem generally associated with wealthy, developed nations and wealth-pursuing developing nations. It is no coincidence that the problem made its first large-scale appearance, and was first recorded and researched, in the heartland of the industrial revolution—England. A wide variety of processes, such as power production, chemical refining, cement mixing, smelting, vehicle use, and biomass burning, contribute to the acid deposition problem. But of all these processes the principal contributors are those related to energy production, in particular energy production in coal-fired power plants and energy production by petroleum-powered vehicles. On a global scale coal-fired power plants are far and away the largest contributor to SO_2 emissions, and are a close second in NO_x emissions; and motor vehicle emissions are the largest contributor to NO_x emissions.

Acid deposition originated as a local air pollution problem. From local origins it grew into a regional- and continental-scale problem. Acid deposition has not yet achieved the dubious distinction of becoming a global problem in the same sense as global climate change or destruction of the ozone layer; however, it may be regarded as global or emerging global in the following senses. First, it is appearing in more and more locations around the globe. As industrialization spreads, the acid deposition problem is similarly likely to spread. If industrialization becomes a global phenomenon, then acid deposition may also become a global problem. Second, acid deposition represents a major human-induced disturbance in global biogeochemical cycles, particularly the sulfur cycle. There is no question that at some level the effects of the disturbance reverberate everywhere on the globe, though they may not be readily detectable and may never precipitate an ecological or political problem. And third, there is already evidence of trans-Atlantic and trans-Pacific transport of acidic substances, and their transport to the Arctic polar region.[2] On a regional/continental scale, there are currently three acid deposition "hot spots" in the world—northern and eastern Europe, eastern United States and Canada, and East Asia.[3]

Acid deposition is a relatively old air pollution problem. Decades of research and policy experiments have resulted in the creation of a relatively complete, yet evolving, standard picture or model of the scientific causes and effects of acid deposition phenomena (a standard acid deposition problem-framework), together with an implied set of generally applicable policy choices for controlling the problem. There are three basic elements of the acid deposition problem—(1) emission of acidic pollutants into the atmosphere, (2) atmospheric transport, transformation, and deposition of emitted pollutants and their by-products, and (3) adverse impacts to humans and ecosystems caused by deposition of the pollutants and/or their by-products. As a result of the problem, policy choices designed to control, prevent, or mitigate the adverse impacts are made (see figure 3.1 later in this section).

The standard acid deposition problem-framework is made up of generic, or universal, methods and knowledge that are theoretically applicable to an acid deposition problem no matter where it manifests—Europe, North America, East Asia, or anywhere else on Earth. Elements of the standard acid deposition problem-framework, when transmitted across boundaries and applied to new locations, are the raw material from which geographically specific (or localized) problem-frameworks are generated. The standard framework contains universal methodologies related to monitoring, emission inventories, atmospheric computer models, and human and ecological impact research, and universal knowledge related to the cause-and-effect mechanisms of the acid deposition problem. As a whole it constitutes a guiding image for scientific and political understanding of the problem. It was first crystallized in the 1970s, and has since become increasingly complex and increasingly integrated with other environmental problems, especially since 1990. The framework's universally applicable methodologies and knowledge are conveyed between scientists by means of science-to-science bridging objects that facilitate creation of localized acid deposition problem-frameworks. From these localized frameworks science-to-policy bridging objects are constructed that can shape policymaking in the region of localization. Early renditions of the standard acid deposition problem-framework were carried to Japan beginning in the late 1970s and early 1980s.

One knows a standard problem-framework is taking hold when textbooks, popular books, introductory pamphlets, and so on treat the problem addressed by the framework in a similar manner and as a coherent whole. There are literally thousands of works published since 1970 that address the acid deposition problem. The images most of them use to describe the problem take their cue from the standard problem-framework, and overall, their structural and textual resemblance is striking. Individual works, of course, vary but their overall pattern is that established by the standard problem-framework. Examples of publications that introduce the framework include: Bubenick (1984), Gould (1985), Seinfeld (1986, 695–728), Mohnen (1988), Pringle (1988), Regens and Rycroft (1988), McCormick (1989), Longhurst (1990), Jacob (1991), Forster (1993), and Howells (1995), to name a few.

The sustainability solution path of the standard acid deposition problem-framework can be encapsulated as follows: to prevent adverse impacts, emissions into the atmosphere must be controlled to a level below which damage is acceptable. This is not quite as trivial as it sounds. It implies that control of atmospheric processes and mitigation of adverse effects after they have occurred are not fruitful avenues for solving acid deposition problems; and it implies that levels of acceptable damage must be determined. This last item is no easy task. The policy component in figure 3.1 represents the political choices that operationalize specific solutions (such as controlling emissions from coal-fired power plants or automobiles) suggested by the general solution path.

Nature and Culture in Japan

Japan's natural environment is where Japan's acid deposition problems play their nefarious games. It is the specific site for locally adapting the generic elements of acid deposition science. A brief introduction to Japan's geography and natural environment is presented in this section and aspects that are pertinent to adapting acid deposition science to Japan are highlighted. This overview should give the reader an indication of the magnitude and complexity of the task of creating a Japan-specific acid deposition problem-framework. In addition, several nature-related aspects of Japanese culture that have been influential in acid deposition science and politics are also highlighted.[4]

The Land

Japan is a long, narrow archipelago dangling off the eastern fringe of the Eurasian continent.[5] It extends approximately 3,000 kilometers (km) in a northeast to southwest direction from the sub-Arctic north (approximately 45°N), through a temperate middle zone, to the subtropical south (approximately 20°N)—roughly the distance and latitude from southern Florida to the Maine/Canadian border. Japan's land area is about 378,000 square kilometers, roughly the size of California, or Finland, or New Zealand. There are four main islands—Honshū, the largest, extending 1,200 km and containing about 60% of the total land area; Hokkaidō, the northern-most island, containing about 20% of the land

	Acid deposition as a human-induced environmental problem originates primarily from the demand in industrial societies for concentrated energy, particularly electricity and motor fuels produced from coal and oil. ⇓
Emissions	Combustion of fossil fuels causes the release of sulfur dioxide (SO_2) and nitrogen oxides (NO_x) into the atmosphere. ⇓
Atmospheric Processes	
Chemical Transformation	SO_2 and NO_x undergo transformations in the atmosphere resulting in formation of sulfuric acid (H_2SO_4) and nitric acid (HNO_3). ⇓
Transport	SO_2, NO_x, H_2SO_4, HNO_3 are turbulently dispersed and carried by winds short and long distances. ⇓
Deposition	and deposited by means of dry or wet processes. ⇓
Impacts	Once deposited they can cause damage to soil, vegetation, inland waters, human health, and human-made artifacts and materials. ⇓
Policy Choices	Recognition of actual or potential adverse effects triggers public policy to alleviate the problem. These policies currently focus primarily on emission control technologies.

Figure 3.1
Major elements of the acid deposition problem

area; Kyūshū, the southern most island with 10% of the land area; and Shikoku, forming the Seto Inland Sea with Honshū and containing 5% of the land area. These four islands constitute about 95% of the total land area of Japan, with the rest made up of almost 4,000 smaller islands.

Japan is surrounded by seas. To the east lies the vast Pacific, to the north the Sea of Okhotsk, to the west the Sea of Japan, and to the south the East China Sea. The latter three seas separate Japan from mainland Asia. These seas not only provide moisture for Japan's abundant precipitation, but also impart to the moisture a chemical composition that is quite unlike that generally found in Europe and North America. The contribution of seasalt sulfate and chloride ions, for instance, is much higher in Japanese precipitation than the precipitation of acid deposition-affected areas in the West. The seas complicate acid deposition research in numerous ways. There are, for instance, relatively few mid-ocean meteorological observations. Yet such data are critical for accurate long-range transport modeling of sulfur and nitrogen oxides. Also, there are few open ocean measurements of dry and wet deposition rates (shipboard measurements of such rates are notoriously difficult to execute). Yet such data are critical for determining the amount of acidic substances reaching Japan from the mainland.

Japan is a land shaped by plate tectonic forces. It is part of a long island arc situated at the eastern edge of the giant Eurasian Plate (which forms the Eurasian landmass) and the western edge of the equally giant Pacific Plate (which occupies most of the bed of the Pacific Ocean and which is moving westward at a speed of several centimeters per year). The Pacific Plate subducts the Eurasian Plate where Japan sits, and it is crustal movements and volcanic activity along this subduction zone that are primarily responsible for not only creating, but also continually reshaping, the islands of present-day Japan. Submarine volcanoes from an era before the Japanese landmass emerged from the seas created an abundance of rich veins of precious metals—gold, silver, and copper. Not only is there an abundance of such metals, but historically they were found in relatively easy-to-work deposits. Thus, Japan was famous in ancient times for its metallic riches.

Though Marco Polo referred to Japan as a land of gold in his *Description of the World* published about 1300, in actuality it was not so much gold as copper for which Japan was famous. Copper was a major export item of Japan from ancient times until almost the present. (It is no accident that the largest bronze statue in the world, constructed over 1,200 years ago, is located in Japan.) Copper is relevant to our history of acid deposition science and politics because the first instances of widespread acid deposition damage in early modern Japanese history were due to copper mining activities around the turn of the century. Indeed, Japan's first environmental crisis was precipitated by copper mining.

Japan is a land of mountains. Mountains cover about 75% of the total land area. The rest is made up of plains (13%) and plateaus (12%). Mountains are the result of the above-mentioned plate tectonic forces. Flat land is rare in Japan. Most plain areas are found along the coast, or inland along the larger rivers. Because of the steep, rugged topography only about 17% of the land area is suitable for cultivation. The two largest plains are the Kantō Plain around Tokyo and the Kinki Plain around Osaka. Much of the population of Japan is concentrated in these two flats.

The checkerboard of mountains that fills Japan's landscape has created pockets of high acid deposition. Some of the main pockets include the Kantō Plain, the area around the Inland Sea, the Sea of Japan coast facing the Asian mainland, and the area around Sakurajima volcano in southern Japan. Like the seas, the mountainous topography presents serious obstacles and major challenges to acid deposition research. For example, it complicates accurate computer simulation because of the difficulty in characterizing precipitation formation mechanisms, wind fields, surface roughness, and so forth in mountainous areas. Computer simulation in Japan is, therefore, more daunting than for the flatter terrain of western Europe or eastern North America.

Japan is a land of volcanoes. There are some sixty-seven active and seventy-seven extinct volcanoes in Japan (constituting about 10% of the world's volcanoes). These volcanoes are part of the so-called circum-Pacific volcanic zone, or "Pacific Rim of Fire." Intense geothermal activity associated with the volcanoes has produced Japan's numerous and famed hot springs. The most famous volcano is, of course, 3,776-meter Mt. Fuji. The most active volcano today is Sakurajima in southern Kyūshū. It almost continuously emits clouds of gas and ash, punctuated with small explosions, and it is the largest single volcanic emission source of SO_2 in East Asia. Unlike western Europe and eastern North America, volcanoes play a major role in the acid deposition problem in East Asia. Volcanic background emissions are almost completely absent from the acid deposition-impacted areas of the West. It has been estimated that total volcanic emissions of SO_2 in Japan rival total domestic emissions (see the appendix for further details). Thus, to precisely estimate anthropogenic contributions to acid deposition in Japan, the contribution from volcanic emissions must be subtracted out. However, volcanic emissions are hard to measure and their trajectories hard to follow. Here again is another challenge to creating Japan's acid deposition problem-framework.

Japan is a land of acidified environments.[6] Besides emitting acidic gases, Japan's volcanic activity has created a significant number of acidified environments—acidic soils, acidic hot springs, and acidic streams and lakes. In these environments a wide variety of acidotolerant and acidophilic life-forms have evolved. Acidic lakes and rivers in Japan are almost all due to acidic waters of volcanic origin. Many have long histories. The most acidic lake in Japan is Yūgama in the crater of Mt. Shirane in Gunma Prefecture in central Japan with a pH of 1.0. No life exists in this lake. However, in another lake with a pH of 1.8, Lake Katanuma in Miyagi Prefecture in northern Japan, there is a wide variety of simple life-forms—green algae, diatoms, plankton, and insects, forming a simple food chain. Lake Katanuma was discovered in 1932 and was at the time of its discovery the most acidic lake found anywhere in the world. Lake Usori (alternatively called Lake Osorensan) in Aomori Prefecture in far northern Japan has a pH of about 3.6, and compared to Lake Katanuma there is an even more complicated food chain topped with a few fish species. This food chain evolved naturally over the 20,000-year history of the lake. In addition to aquatic species, terrestrial species in Japan (grasses, trees, insects, animals, etc.) have evolved to thrive in the highly acidic soils associated with volcanic areas. For example, Oshima cherry trees found on volcanic Oshima Island in the Pacific Ocean south of Tokyo are noted for their acidotolerance. In summary, acidified environments endowed Japan with an abundance of acid-tolerant life-forms, which in turn provide scientists with natural laboratories for the study of the evolution of lifeforms in highly acidic conditions. Research on these ecosystems offers insights into the rehabilitation of lakes and streams irredeemably acidified by human hand.

Japan is an unglaciated land, and therefore not a land and culture of lakes. There are numerous beautiful and unusual lakes in Japan, but because of the lack of glacial scrapping the topography remains steep and lakes few. Therefore, lakes are not a dominant feature of the Japanese landscape as they are of the acid deposition-impacted areas in southern Scandinavia and eastern North America. Therefore, too, the type of highly acid-sensitive ecosystems found in these areas are generally absent in Japan. In Europe and North America glaciers scoured the land, leaving behind exposed or shallow (and nonchemically reacting, non-acid neutralizing) granite bedrock, thin soils, and thousands of small, poorly buffered lakes. In short, they left behind large areas highly vulnerable to acidic inputs. It is these areas where negative aquatic impacts (e.g., damaged fish populations) are most prominent. Japan contains no such equivalently impacted areas.

Many of the areas hard hit by acid deposition in the West can be called areas with a *lake culture*. They are lake cultures not only because of the sheer number of glacial lakes but also because of the role these lakes played in the histories of the peoples who lived there. To give but one example, Canada's inland lakes served as water highways for the original native peoples (the famous birchbark canoe was invented in North America) and for early trappers, traders, and explorers. Japan, by contrast, is definitely not a lake culture. As compared to the saltwater seas surrounding Japan, freshwater lakes have played only a minor role in shaping the Japanese way of life.

The Climate

Japan is located in the temperate monsoon climatic region of East Asia.[7] Following the typical monsoon pattern, winds blow from the continent to sea in winter and from the sea to continent in summer. In winter, Eurasian continental influences dominate. Cold winds whistle eastward off the continent (carrying, among other things, pollutants from northern China, eastern Russia, and the Koreas), and sweep over Japan on their way to decay in the mid-Pacific. In summer, the influence is opposite. Pacific Ocean tropical maritime influences dominate, and warm winds blow northward from the South Pacific (carrying, among other things, pollutants from southern China and Taiwan to Japan, and from Japan to the mainland).

Japan's monsoon climate brings abundant rainfall. Abundant rainfall is one of the defining nature-related elements of Japanese culture. Indeed, Japan can be called a culture of rain. The millennia-old association with rain imparts to the "acid rain" issue an intimacy that other atmospheric issues such as global climate change and stratospheric ozone depletion do not have. The annual average rainfall in Japan is about 1,800 millimeters (mm). This is greater than the rainfall of the other two major acid deposition hot spots of the world—northern Europe (600–900 mm), and eastern North America (about 800 mm). Interviewees of all stripes repeatedly stressed to me that rain is special to the Japanese people. It is strongly affiliated with the long and honorable history of the nation and with the Japanese spirit. In a sense, to contaminate rain is to contaminate their history and spirit. Thus, for many political actors, especially the general public, the acid rain issue tugs at cultural heartstrings (and intensifies the political importance of certain bridging objects). For these actors, the purity of rain, because of its agricultural, historical, aesthetic, spiritual, and other associations, must be preserved. Public pressure on the government to protect the rain began to materialize in the early 1990s when a "pH monitoring craze" hit Japan.

Japan has four distinct seasons. The four seasons can be fairly neatly partitioned as follows: winter (December, January, February), spring (March, April, May), summer (June, July, August), and autumn (September, October, November). Each is briefly explained here for each is relevant to Japan's acid deposition history and to Japan's localized problem-frameworks.

Winter begins with the arrival of cold northwest winds from the continent around the end of November. The winds originate in a massive high pressure system that builds during the winter in the interior of the Eurasian continent over Siberia and Mongolia, and it is this high pressure system, called the "Siberian High," which dominates Japan's winters from the end of November to the end of February when the high recedes. For three months it almost continuously sends cold northwest winds over

Japan. Air coming from the Siberian High is initially very cold and dry over the continent. However, on passing over the Sea of Japan it absorbs heat and moisture. Much of the heat and moisture is provided by the warm Tsushima current from the southern seas.

When the newly moisture-laden air hits the mountains of the Sea of Japan coast it drops its load and brings a veil of clouds and deep snowfall. Hence, the Sea of Japan provinces have traditionally been referred to as *yukiguni*, or snow country. The air, now emptied and dry, brings clear, sunny skies to the Pacific coast side of Japan, though there are occasional snowfalls. Snow, however, is rare in Kyūshū, and never falls on the Ryūkyū islands in the far south. Japan's winter weather is often referred to as the "winter monsoon." Air pollutants from the Asian mainland ride the winter monsoon winds and bring the largest concentrations of acidic pollutants from the mainland of any season. These pollutants fall mainly on the Sea of Japan coastal region, and in some areas are estimated to contribute up to 50% of total acidic deposition (see the appendix). This long-range transport of acidic pollutants was first detected in the mid-1980s. Today Japan's "victim" status as a net importer of pollutants motivates its international policymaking on the transboundary air pollution issue in East Asia.

Spring traditionally begins with the flowering of white plum blossoms in March. This roughly corresponds to the decay of the Siberian High. Temperatures increase on the Asian mainland, and southern maritime lows begin to develop. The lows as they move north bring warm winds and cause a rapid rise in temperature. The true arrival of spring is announced by Japan's proverbial cherry blossoms in April. April is followed by a month of fine, stable May weather, and May yields to Japan's "gloomy" rainy season in June.

April also marks the appearance of a phenomenon known in Japanese as *kōsa*, or yellow sands. During this month winds originating in the deserts of Mongolia and the arid regions of northern China carry yellow-colored loess dust that can tinge overcast skies yellow.[8] These dust particles are alkaline and play a major role in reducing the impacts of acid deposition in East Asia because they neutralize a significant fraction of East Asia's acidic substances. The effect is most pronounced in northern China, but also applies to the Korean Peninsula and Japan.

Up to 20–30% of Japan's acids are estimated to be neutralized by the yellow sands. *Kōsa* is carried on winds to Japan primarily in the spring, around April, and during this period can neutralize acids in the atmosphere (a phenomenon that must be considered in long-range transport computer models). However, once deposited on soils the alkaline substances can also neutralize acid inputs throughout the year. *Kōsa* deposition is not uniform over Japan, and this fact must be taken into account in ecological impact research. Without the fortuitous occurrence of the yellow sands phenomenon, the acid deposition problem today would be far worse in Japan, to say nothing of the Koreas and northern China.

Summer begins with the rainy season known as *baiu*, or plum rains,[9] which in Japan lasts for about 30 days from mid-June to mid-July. The plum rains are a major feature of the climate of East Asia. They are the product of the *baiu* front and are essential for the early stages of rice cultivation throughout East Asia. The front forms in southern China in May and gradually moves northward passing over Japan. It is characterized by heavy overcast skies, light winds, high humidity, frequent gentle rains, and now by pollutants from Taiwan and southern China. It was during the rainy seasons of the early 1970s that one variant of Japan's acid deposition problems (called "moist air pollution") was discovered. This prompted the first major, organized acid deposition research program in Japan.

After the *baiu* comes a short, dry period in midsummer. This is a lull that extends from roughly mid-July through the end of August and separates the early summer rainy season from the autumn rainy season. The skies are clear, but the weather hot and oppressive, almost tropical in character. There are occasional deafening thunderstorms. Summer is the time of the largest amount of acid deposition produced by domestic sources. This occurs especially around the large urban centers. Because of the high temperatures and humidity the problem is especially acute for those acid deposition-related chemical processes associated with photochemical reactions.

Autumn begins with a bang—typhoons—most of which occur in September. As the North Pacific High recedes, cold air masses in the north move southward, producing a front known as the "*shūrin*, or autumn rain" front. This change causes typhoons to develop in the South Pacific that often pass along the *shūrin* front, bringing torrential rains and pummeling winds. By mid-October, however, fine, dry weather returns. Stable air produces a brilliant blue "high" sky. Fall colors begin to appear in October just as crops are being harvested. Autumn is Japan's mildest acid deposition season. Deposition values are the lowest of any of the four seasons. However, in far off central Asia the Siberian High is beginning to build again.

The Soil and Vegetation

Japan is a land of complex patterns of bedrock and soil distribution. Because of its location in a highly active plate tectonic zone, and because of its mountainous, abundant rainfall, and nonglaciated character, the distribution of underlying bedrock and overlying soil is highly complicated (Katoh et al. 1990, 46–48). This is unlike the acid deposition-impacted areas of Europe and North America that tend to have relatively uniform bedrock and soil types by comparison. This influences the nature of the acid deposition problem in Japan in a variety of ways. As one example, Japan's many-branched rivers tend to run through several geological zones, some of which inevitably contribute alkalinity to the waters. The end result is that almost all of Japan's rivers are well buffered. Those that are not tend to be acidified due to volcanic activity. The complicated geology and pedology complicates impact-related acid deposition research. There are few neat blocks of land with consistent bedrock, soil, and vegetation type.

Japan is a land and culture of forests. Japan's diversity of climatic zones historically produced an amazing variety of ecological niches for a wide diversity of vegetation. The most conspicuous vegetation type is broadleaf and evergreen temperate forests. About 70% of the total land area of Japan is covered in forest, a very high percent for an advanced industrial nation. Japan has almost 200 varieties of trees. They range from subarctic spruce and fir in Hokkaidō to palms in the Ryūkyū Islands. The three most common trees are the Japanese cedar (*sugi*), Japanese cypress (*hinoki*), and Japanese red pine (*akamatsu*). Because of its ubiquitous forests and traditional reliance on forest products, Japan can be called a "culture of wood." Plentiful wood, and earthquakes, made Japan a culture of wood.

In its early days of nation-building, Japan attempted to become a culture of stone. Stoneworking techniques, as many other aspects of early

Japanese civilization, were brought over from the Korean Peninsula. In the Asuka area near Nara (the seat of the Yamato court during the sixth and seventh centuries) one can still see today the remains of a variety of stone structures. They are carbon copies of stonework found in Korea. But by the time of the establishment of the first capital at Nara in 710, large-scale construction in stone had almost completely disappeared. The reason was simple. The stonework of that day and age could not stand up to Japan's frequent earthquakes. (By contrast, the Korean Peninsula is virtually earthquake free.) Thus, just as the Japanese state reached the point it could build monumental architecture, it chose wood, not stone, as its primary material.

It is therefore not surprising that the oldest wooden structure in the world, the Buddhist temple Hōryū-ji, is found in Japan (in Nara). It dates from about 670. In addition, the largest wooden structure in the world, the Buddhist temple Tōdai-ji, is also found in Nara. It houses the largest bronze statue in the world, and dates from around 750. Besides wooden structures, wooden items also permeate traditional culture—chopsticks, rice paddles, ear cleaners, musical instruments, rice warmers, elegant tea whisks, water buckets, *sake* casks, palanquins, and so on. Given the practical, intimate, and millennial-old attachment to forest products in the Japanese culture, it is not surprising that threats to "wood" (forests, trees, wooden structures), as they are to rain, touch deep cultural heartstrings. Again, public pressure on the government to protect forests from acid deposition began to form in the early 1990s.

The preceding description of Japan's geographic setting characterizes some of the natural and cultural contexts for Japan's confrontation with acid deposition problems. We will revisit many of these contexts during our inquiry into Japan's acid deposition history. We turn now, though, to a brief examination of one vital, nontraditional component of Japanese society that permeates our study—science.

Science in Japan

Japan today, like other advanced industrial nations, has numerous and sophisticated environmental scientific communities that produce numerous and sophisticated scientific data on a wide variety of environmental problems. The full story of how Japan came to this highly evolved state, though fascinating, is beyond the scope of this book. However, as background to our acid deposition history, the development of science in Japan is briefly introduced here. Japan's mastery of modern science (and technology) is a remarkable achievement. It was the first non-Western nation to fully adopt and apply modern science. Thus, it is understandable that much of the literature on the history of science in Japan is preoccupied with fathoming how a culture with ways of thinking seemingly incompatible with those of Western civilization could in such a short time around the turn of the twentieth century become a scientific powerhouse. Most scholarly research focuses on the question: How and why did Japan, after around 1850, build a tradition of scientific research out of seemingly nothing?

Works in English that treat the history of science in Japan include the following: Sugimoto and Swain's (1978) treatise is the most complete account in English of the history before 1854. Shigeru, Swain, and Eri (1974) explore multiple facets of the post-1854 history in a comprehensive edited volume. The fullest exposition in English of the social history of modern Japanese science from 1868 to 1920, the period during which Japan's tradition of scientific research was created, is contained in Bartholomew (1989). Case studies by Watanabe (1990, 1997) add to the English literature. Further information can be found in the journal *Japanese Studies in the History of Science.*

From the dawn of Japanese civilization to 1900, Japan twice completely overhauled—or modernized—its entire social system. During both periods, "importation of new knowledge and the creation of new systems for its advancement and use in society were at the heart of the process" (Sugimoto and Swain 1978, 1). The first overhaul occurred slowly between the years 600 and 900 as Japan adopted the Chinese social model of the Tang period, and the second occurred rapidly between 1854 and about 1900 as Japan frantically absorbed Western knowledge. During and between these times of transformation, Sugimoto and Swain (1978) identify five periods or waves during which "science" (by which they mean "systematized knowledge") was imported from abroad: (1) Chinese Wave I (ca. 600–894), associated with assimilation of the luminescent knowledge of Tang dynasty China; (2) Chinese Wave II (1401–1854), associated with a more sophisticated retention of Confucianism and practical Chinese learning than during Wave I; (3) Western Wave I (1543–1639), associated with the Portuguese traders and Jesuit missionaries who first brought Western knowledge and technology to Japan, and which was cut short with the imposition of a policy of national isolation in 1639; (4) Western Wave II (1720–1854), associated with *Rangaku* (Dutch learning), dating from the time when the Dutch were allowed to establish a trading post near Nagasaki, thus providing a tiny window for Western knowledge to seep into Japan; and (5) Western Wave III (1854–present). Sugimoto and Swain postulate that the first four waves set the stage in various direct and indirect ways for the rapid assimilation of Western science after 1854.

Though the literature on Japan's modern history of science is sizeable, I am unaware of any works in either English or Japanese that address in more than cursory fashion the history of environmental science, let alone the history of environmental science and politics. Shortly though, we will be embarking on a journey through one of these histories, the history of Japan's acid deposition science and politics. However, before we do so I want to briefly address a question that underlies this history: *Does Japan's culture affect its practice of science?*

Science is reputed to be a universal institution, and the mistaken assumption is often made that all scientists, no matter where they practice science, do so in the same way. Japanese scientists, however, do not practice science exactly like their counterparts in the West or elsewhere. In many ways, science is an ill fit with *traditional* Japanese culture. Science is an imported way of thinking overlaid on an ancient culture that lacks many of the basic intellectual premises on which science was founded in the West. This breeds a fundamental tension between the conduct of science in Japan and some traditional modes of operating within society. The tension is often difficult for non-Japanese to appreciate, but it is what makes Japanese scientific practice distinct from Western practice.[10] The conclusions that follow are my own; however, they were developed and enhanced from comments gathered during interviews and discussions with Japanese researchers.

First, among Japanese scientists, and compared to scientists in the West, there is a general lack of open questioning. In Japanese society there is less use of "why" questions because they are often considered intrusive. Science as it evolved in the West is premised on a perpetual asking of why questions. This is what underlies the ancient dictum of Western science to "question all authority," whether Greek philosophers or the Bible. On a personal level, asking "why" questions in Japan can cause a person to lose face, placing them in an awkward situation, or causing them embarrassment, or forcing them to express an opinion they would rather keep to themselves. On a societal level, asking questions can lead to political disruption, which perturbs the harmony of society. Thus, no matter what the social level, Japanese individuals traditionally learn from an early age to suppress aggressive questioning. This carries over into science and into acid deposition science. Scientists may be reluctant to question another scientist's methods or data or conclusions. Junior scientists, for instance, may be reluctant to question senior scientists' results even if they are flawed. The inhibition to open questioning can lead to harmony among scientists, but also at times to sloppiness in scientific research.

Second, and related, there is a tendency to avoid vigorous and controversial discussion in the Japanese scientific world. Japanese society tends to value silence over conversation, especially when the conversation may be controversial. Debate, for instance, has never been a prominent part of a Japanese scholar's (or politician's) repertoire as it has in the West. East Asian societies in general have little tradition of debate. Confucianism, for instance, did not glorify debate or public speaking as a skill to be cultivated. By contrast, the "silent arts" of calligraphy, poetry, and painting were highly valued. There are precious few great public speakers in East Asian history like Athens's Solon or Rome's Cicero. The emphasis on "silence is golden" in Japan leads to a lack of friction among scientists, but also at times to a lack of intense verbal interchange. As a personal observation, I was deeply impressed by the courteousness and attentiveness with which audiences generally listened to speakers at conferences, meetings, and small group seminars. I was equally impressed by the lack of lively discussion, debate, or exchange of ideas.

Third, Japanese scientists generally do not like to take professional or intellectual risks. This inhibits development of activist (versus inactive or active) members of expert communities. Japanese culture does not enshrine the individual or individual display, and thus there is a weak or nonexistent system of rewards for successful risk-takers. There is a quiet caution within the Japanese scientific world that runs counter to the celebration of individualism in Western science. Published scientific conclusions, for instance, are often understated, indirectly stated, or just allowed to naturally bubble to the surface without comment—the no-conclusion conclusion. As products of a nonindividualistic society, many Japanese researchers are repelled by the cutthroat competition that permeates much of the Western scientific world, often lamenting that Western researchers too often seem preoccupied with their ego and the race to be first.

Fourth, there is often an insularity to Japanese thinking (scientific and nonscientific). Ultimately, this stems from the fact that Japanese society evolved on an isolated set of islands. Japanese scientists, for instance, sometimes do not feel the need to share their results with the wider world. This parallels Japan's historic tendency to engage in one-way absorption of ideas from the outside. The Japanese pattern has been to import ideas from foreign cultures, make the ideas their own, and then internally refine them. This first occurred in the Nara period (710–794). It occurred again during the Meiji period (1868–1912). And, as we will see, it has occurred in the acid deposition field from 1868 to the present. Until recently there was not a reciprocal commitment to tell the Western world what Japan was doing in the acid deposition field.

Fifth, there is a strong mentality of "thinking vertically" in Japan. The Japanese often refer to their society as *tateshakai* (vertical society). This means that lines of responsibility, duty, and communication are structured in a vertical, hierarchical pattern. Thus, the Japanese are acutely aware of who is above and below them, and their duties and obligations in relation to these superiors and inferiors, but far less aware or interested in "horizontal" relations. There are many merits to this system, but one defect apparent in the acid deposition field is the difficulty of doing interagency research. Each bureaucracy has its own research territory, and foreign scientists (i.e., those not associated with the bureaucracy) who cross boundaries in the name of integrated research are usually driven back to their home territory. In my interviews and discussions with scientists, many lamented that they could not venture into this or that scientific territory because it was not within the jurisdiction of their home institution. The vertical structuring to scientist-bureaucracy interaction means that Japan's nationwide acid deposition expert community is also a mosaic of subcommunities, not a homogeneous entity. The most important split is along ministerial lines, particularly between the Ministry of Environment (formerly the Environment Agency) and the Ministry of Economy, Trade, and Industry (formerly the Ministry of International Trade and Industry). Each ministerial subcommunity consists of those researchers who receive the bulk of their funds from that ministry. In turn they are loyal to the ministry.

Bureaucrats typically sit higher on the power hierarchy than scientists. Thus, there is a danger that bureaucratic pressures can override a free and objective scientific inquiry because the flow of information is often subject to bureaucratic involvement. Scientists, especially at national and local research institutes, often pass their information first to bureaucrats and only afterwards, if at all, to the mass media, general public, or even peer-reviewed journals. This means that bureaucrats are earlier and better informed than other parties, and can therefore exert considerable control over sensitive scientific information if they deem the situation warrants it. The intimate scientist-bureaucrat relationship makes highly visible fields within environmental science in Japan somewhat high risk. In other words, there is a much higher probability of research results in these fields getting a scientist into trouble with higher officials than there is in, say, the domains of computer chip technology or volcanology.

Generally, the risk diminishes the further a scientist is from the centers of power. Scientists have greater freedom to act, speak, and publish the more distant they are from the center, but the price is often little influence in policy and scarce funding. The omnipresent bureaucratic ax over a scientist's head sometimes has the subtle (and sometimes not so subtle) tendency to steer researchers' attention to bureaucratic correctness instead of, say, environmental protection.

And finally, Japanese culture is strongly group-oriented. This not only fortifies community feeling among the members of Japan's acid deposition expert community, but also means that at times considerations of social harmony dominate other considerations, even the objective pursuit of science. There is powerful pressure to conform in Japan. Of all traditional belief systems, Confucianism is perhaps the most influential on the acid deposition expert community. Confucianism is a potent society-oriented (versus nature-oriented) belief system. Thus, members may sometimes exhibit a stronger tendency to "defend society" (and conventional social norms) rather than "defend nature" (and more recent ecologically based norms).

What is the overall effect of traditional culture on the practice of science in Japan? Although I cannot give a definitive answer, my impression from studying the acid deposition case is that once a sufficient number of scientists are engaged in analyzing a problem and once the science matures to a level such that a solid problem-framework emerges, cultural influences of the sort described here tend to recede to the background. Thus, in my estimation, cultural differences, while significantly informing social interaction among scientists at the microlevel, do not at the macrolevel greatly affect the scientific integrity of Japan's present-day acid deposition problem-framework, nor significantly inhibit and distort the production of bridging objects. However, they do seem to affect the proactive transmission of bridging objects to the policy world.

Japan's Acid Deposition History

The history of Japan's acid deposition science and politics is intertwined with two larger domestic histories—Japan's experience with environment problems in general, and air pollution problems in particular. Great pains are taken in this book to situate Japan's acid deposition history within its air pollution and environmental histories. On occasion we will digress into these histories to highlight certain trends and characteristics that are relevant to the history of the acid deposition problem. This section presents the chronology of Japan's intertwined environmental and acid deposition histories used in this book. The swath of history from 1868 to the present is divided into five environmental eras, and six periods related to acid deposition problems. There is a large, though still underdeveloped, body of literature that I drew upon for my understanding of large-scale trends in Japan's environmental history.[11] I divide the environmental history of Japan into five eras, as follows:[12]

1. 1868–1945: Prewar Industrialization Era
2. 1945–1967: Postwar Reconstruction Era
3. 1967–1973: Domestic Environmental Revolution
4. 1973–1990: Transition Era
5. 1990–present: Global Environment Era

The roughly 125-year span between 1868 and the present is partitioned into the following six acid deposition periods:

1. 1868–1920: Copper Mines and Early Precipitation Chemistry
2. 1920–1945: Flowering of Precipitation Chemistry
3. 1945–1974: Reign of Air Pollution
4. 1974–1983: Moist Air Pollution
5. 1983–1990: Nationwide, Ecological Acid Deposition Research
6. 1990–present: East Asian Transboundary Air Pollution

The transition from one acid deposition period to another is marked by an epistemic shift from the prevailing problem-framework to a new one. Indeed, identifying a shift allowed me to distinguish the periods. I identified five such shifts between 1868 and the present. In the succeeding chapters they are explained and the evolution of Japan's acid deposition science and politics during each of the six periods is traced. Essentially, the six periods constitute six case studies. Comparing each for what components are present and, very importantly, what are missing in terms of problem-frameworks, expert communities, bridging objects, etc. gives leverage for understanding how science and policy interact in general.

We begin our history with the first seeds sown soon after the Meiji Restoration of 1868. For each period, following fairly strict chronological order, I present the relevant published literature, historical incidents, and policy decisions without much in the way of analytical comment. Then at the end of each chapter describing a period, the main features are summarized and analyzed.

No systematic and comprehensive history of acid deposition-related science and policy in Japan exists. However, a number of articles provide sketchy histories. Some of these include: Okita (1983), Fujita (1987), Katoh et al. (1990), Hara (1993), Maruyama et al. (1993), and Fujita (1993a).[13] I relied on these articles to initiate my historical research; however, the vast bulk of information pertaining to Japan's acid deposition science and politics presented in this book derives from original sources. In addition, the present state of acid deposition science and politics is largely based on formal interviews with over fifty scientists and policymakers, informal discussions with many others, and my own personal participation in a few of the events described, besides the available scientific literature. Most of the research for this book was undertaken during my two-year (1994–1996) stay at the National Institute for Environmental Studies of the Ministry of Environment of Japan.

The full range of disciples contributing to acid deposition-related knowledge is intimidatingly large. This poses the exasperating problem for a researcher trying to reconstruct the historical buildup of such knowledge of deciding cutoff points between relevant and irrelevant knowledge. To simplify the task I followed a couple of self-devised rules.

The first rule was to focus primarily on research related to precipitation chemistry until such time as acid deposition was recognized as a clear-cut environmental problem in Japan in the mid-1970s. It was not until the mid-1970s that other fields of study were linked to precipitation chemistry with the express objective of unraveling the cause-and-effect mechanisms of the acid deposition problem. Precipitation chemistry is what might be called the "foundation discipline" of acid deposition science. It was precipitation chemistry monitoring, for instance, that led to the discovery of large-scale atmospheric acidification problems in the West. My focus on precipitation chemistry means I ignore many and major historic developments in Japan in the fields of soil and freshwater chemistry, forestry, materials science, atmospheric physics and chemistry, and so forth, at least until they became relevant to the contemporary acid deposition problem. Thus, the early history of acid deposition-related knowledge in Japan as it is portrayed in this book is largely a history of precipitation chemistry, not a comprehensive history of all scientific fields eventually relevant to the acid deposition problem.

Table 3.1
Japan's environmental and acid deposition chronologies

Dates		Period Name	
Environmental History	Acid Deposition History	Environmental History	Acid Deposition History
1868–1945		Era 1: Prewar industrialization	
	1868–1920		Period 1: Copper mines and early precipitation chemistry
	1920–1945		Period 2: Flowering of precipitation chemistry
1945–1967		Era 2: Postwar reconstruction	
1967–1973		Era 3: Domestic environmental revolution	
	1945–1974		Period 3: Reign of air pollution
1973–1990		Era 4: Transition	
	1974–1983		Period 4: Moist air pollution
	1983–1990		Period 5: Nationwide, ecological acid deposition research
1990–present		Era 5: Global environment	
	1990–present		Period 6: East Asian transboundary air pollution

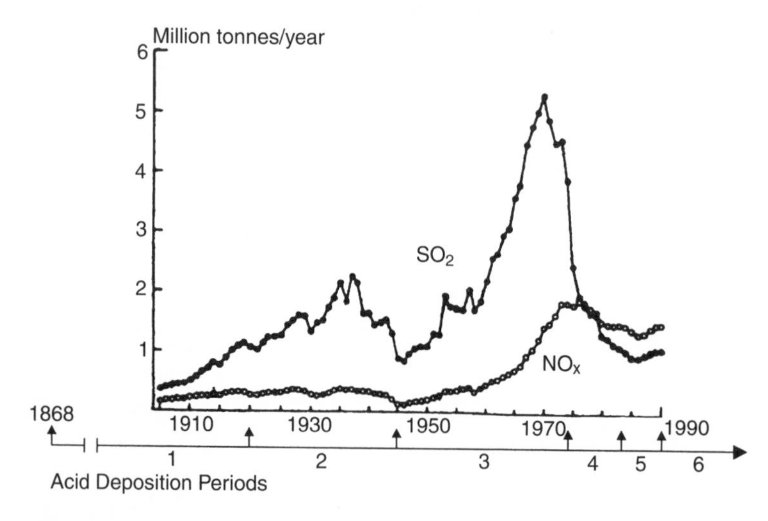

Figure 3.2
Acid deposition periods juxtaposed against SO_2 and NO_x emissions (1900–1990). Reprinted from Shinichi Fujita, "Overview of Acidic Deposition Assessment Program in CRIEPI," in *Proceedings of CRIEPI International Seminar on Transport and Effects of Acidic Substances, November 28–29, 1996*, ed. Yoshihisa Kohno, p. 5 (Tokyo: Central Research Institute for the Electric Power Industry [CRIEPI], 1997). Used with permission of the author.

The second rule I followed was to confine myself "chemically", for the most part, to precipitation chemistry knowledge related to pH, sulfate (SO_4) compounds, and nitrate (NO_3) compounds. Sulfate and nitrate compounds are historically, and still at present, the major contributors to the anthropogenic acid deposition problem. They are, however, by no means the only substances involved in acidification problems. Invoking this chemical rule means that I largely ignored research related to other chemical species. For instance, I generally do not describe research on sulfur dioxide when that research does not specifically address the conversion of SO_2 to SO_4.

The preceding rules did not solve all my problems. Many prewar and even postwar articles are very poorly referenced, for example. Names, dates, sources for data, etc. were often not cited. This sometimes made it difficult to determine exactly where ideas came from. Many times I couldn't, for instance, determine a scientist's source of inspiration or a policymaker's source of information. To the fullest extent possible, I attempted to track the flow of ideas related to acid deposition between the various actors involved; however, there were many times when the trail became dim or disappeared altogether. Despite this, I hope the reader will be able to visualize the overall network of interactions even when individual trails are lost in the fog.

The chronological delineation used in this book to analyze Japan's environmental and acid deposition histories is given in table 3.1. The next six chapters discuss in detail each of the six acid deposition periods. Figure 3.2 shows these acid deposition periods juxtaposed against a graph of estimated annual emissions of sulfur dioxide and nitrogen oxides from 1900 to 1990.

4

Period 1 (1868–1920): Copper Mines and Early Precipitation Chemistry

First Environmental Era (1868–1945): Prewar Industrialization

The entire period from the Meiji Restoration of 1868 to the end of World War II constitutes a single era in Japanese environmental history. However, it constitutes two periods in Japan's acid deposition history. There is a relatively consistent environmental pattern throughout this time span that can be summed up as follows: industrial development reigned supreme; the Japanese people were proud of their development and could (and would) absorb the relatively minor detrimental aspects of industrialization such as pollution-induced illness and degradation of natural surroundings; the environment, likewise, it was assumed, could (and would) absorb all blows delivered by industrial development; and, what conflicts that did arise between industry and citizens over the degradation of the environment could (and would) be dealt with through a nonlegislative, nonjudicial, time-honored personal mediation process. During Japan's first environmental era no national-level environmental legislation was passed. Sporadic citizen movements emerged in response to particularly nasty local pollution problems, but at no time, except for one brief moment around 1900 (which will be discussed in detail later, and which was Japan's first experience with the unsustainability of modern civilization), did these local movements coalesce into a nationwide movement and generate a nationwide environmental consciousness. Industrialization, militarization, and imperialism were the watchwords during this era. Japan's first environmental era, and first acid deposition-related historical period, begin with the Meiji Restoration.[1]

The Meiji Restoration of 1868 marked the end of the 250-year rule of the Tokugawa Shogunate and the emergence of modern Japanese society. The arrival of Commodore Perry's cannon-wielding black ships in the summer of 1853, and the subsequent realization among the elite that Japan lagged far behind the powerful nations of the West, especially in technological development, convinced many of them that Japan needed to break its isolation, open to the West, and embark on a course of rapid industrialization. During the 250 years of the Tokugawa Shogunate, Japan was isolated from the rest of the world to a degree almost unprecedented in world history for a large, well-established culture. The isolation officially ended with the "restoration" of the Emperor Meiji to the position of supreme ruler of the Japanese nation in 1868. During the years preceding and following the Restoration, Japanese leaders initiated a process of rapid technological, social, educational, political, economic, and military overhaul. A feudal, agrarian society was turned almost overnight into a modern industrial nation.

The jump-start to industrialization, however, was not without environmental consequences. Black smoke, offensive odors, periodic epidemics, and general industrial grunge became commonplace, especially in the large cities of Tokyo, Osaka, and Nagoya. In the early years after the Meiji Restoration the rush to industrialization led to numerous environmental pollution incidents, the most prominent of which were caused by copper mines.[2]

Copper Mines

All early, unequivocal instances of acid deposition damage in Japan are related to emissions of sulfur dioxide from copper smelters. The four major copper mines were the Ashio, Besshi, Hitachi, and Kosaka mines. In addition, there were dozens of smaller mines. Major air and water pollution damage, and citizen protest movements, were experienced at most of the mines. The Ashio Copper Mine precipitated Japan's first environmental (and sustainability) crisis. It was also the site of Japan's first major acid deposition problems. While the most significant environmental damage at Ashio was due to water pollution, such was not the case at the other three big copper mines. Environmental damage at each of the Besshi, Hitachi, and Kosaka mines was primarily related to SO_2. The histories of the Ashio, Besshi, and Hitachi mines are discussed in this chapter. Kosaka does not add greatly to our understanding of early acid deposition science and policy in Japan, so its history is not included.

Ashio Copper Mine (1877–1907)

The infamous Ashio Copper Mine Disaster (or *Ashio Dōzan Kōdoku Jiken*—Ashio Copper Mine Public Poisoning Incident, as it is euphemistically referred to in Japanese) was the defining event of Japan's first environmental era. It is the subject of a huge literature.[3] Most of it deals with Ashio's water pollution problems, the aspect that created the disaster. This portion of the story is related only briefly here. Our main focus is on the peripheral acid deposition portion.

The Ashio Copper Mine is located deep in the mountains of the upper reaches of the Watarase River in western Tochigi Prefecture 160 kilometers north of Tokyo. The mine has a long history. It was discovered in 1610. The Tokugawa Shogunate quickly took over control and developed it into a major mining center. By the late 1600s it was the most productive mine in Japan. High-quality copper ore was extracted from *tanuki* (badger) holes hand-hewn into the mountains. Ashio copper was a valued item during the entire Tokugawa or Edo period (1600–1868). It was a prized decorative material used in the Tokugawa shogunate's massive castle in the city of Edo (today's Tokyo) and the ostentatious mausoleum at Nikkō honoring Tokugawa Ieyasu, the founder of the Tokugawa shogunate. It was even an important export commodity in Japan's trickle of trade with China and Holland. However, by the time of the Meiji Restoration the mine was dormant. It became private property in 1871.

Furukawa Ichibei purchased the mine in 1877.[4] This date marks not only the launch of Ashio's meteoric rise to becoming the largest copper mining complex in Japan, and the largest in all of East Asia, but also the beginning of its continuing string of environmental disasters. Furukawa rapidly modernized and scaled up mining operations. In 1884 Ashio's mother lode was discovered. This discovery "began the transformation which was to turn Ashio into one of the world's greatest and most highly modernized copper mines by the turn of the century" (Notehelfer 1975,

356). Between 1880 and 1907 the Ashio mine produced between 20% and 50% of Japan's annual copper output (Shoji and Sugai 1992, 20), and by 1900 it employed almost 20,000 workers. Most of Ashio's, and Japan's, copper was exported. In 1890 copper constituted almost 10% of Japan's export earnings (Shoji and Sugai 1992, 18). Since copper was one of the chief foreign exchange earners, and since Furukawa was the largest owner of copper mines in Japan, this ensured access to, and the support of, the inner circles of political power. When environmental disaster upon disaster began piling up at Ashio, Furukawa's political connections at the highest levels guaranteed favored treatment.

The three major actors in the Ashio drama were the national government, primarily the Ministry of Agriculture and Commerce; the Furukawa Mining Company; and local citizens along the rivers downstream of Ashio. The relationship between government and industry (the Ministry of Agriculture and Commerce, and Furukawa) was intimate. The relationship between the local citizens and Furukawa was adversarial. And the relationship between the local citizens and the government was distant. Science played a modest role in the politics of the crisis.

As mining operations intensified, the Watarase and Tone rivers became seriously contaminated from acidic and heavy metal pollutants. The Watarase River exits the mountains into the flat expanse of the Kantō Plain some 75 kilometers downstream of Ashio and joins the Tone River some 25 kilometers later. The most serious water pollution damage was experienced in the vicinity of the junction of the two rivers, approximately 100 kilometers downstream of mining operations. Historically the region around the Watarase and Tone rivers prospered.

The amounts of mine tailings and smelter slag from the Ashio mine were not large in the Edo period; however, after 1877 the disposal of mine waste became difficult because of the growing volume. The problem was "solved" by dumping the wastes into the Watarase River. A system of disposal developed whereby the wastes were heaped along the riverbanks, and then when heavy rains came, workers were called out to dynamite the heaps and shovel them into the river. This, of course, resulted in massive downstream pollution.

Farming and fishing communities along the Watarase and Tone rivers began to experience signs that all was not well as early as 1878. In 1888 the first major flood occurred, and in 1889 the first major harvest failed. In 1890 a second flood inundated 1,600 hectares of farmland and damaged 28 villages (Shoji and Sugai 1992, 22). After this, floods occurred regularly. In addition, polluted irrigation water drawn from the river further damaged the rice fields, and illness among the people appeared.

The outbreak of the Sino-Japanese War (1894–1895) caused an even higher demand for copper. Floods, harvest failures, and health problems continued unabated. In 1896 the most devastating flood to date hit the area. One large city and 136 villages were damaged, and over 50,000 hectares of farmland were poisoned. Hundreds of people died and more than 500,000 were affected (Shoji and Sugai 1992, 27). A sense of crisis prevailed. Affected villagers decided to bypass the prefectural government and directly petition the national government to close the mine. Four mass demonstrations (marches on Tokyo) took place.

In February 1900 the last and largest demonstration took place when some 3,000 to 12,000 people (estimates vary widely) began a march to Tokyo. Along the way they were intercepted by police, and a full-scale battle ensued. The march was stopped, and the local people seemed to be defeated. However, by this time, Ashio was becoming a national issue. A large support movement slowly emerged, and major newspapers, particularly the *Mainichi* and *Yoruzu* newspapers, widely broadcast news related to Ashio. Inspection tours and lantern slide shows of the devastation and suffering were conducted by religious organizations, women's groups, and student associations.

Japan was now once again in the throes of war preparation (the Russo-Japanese War of 1904–1905). To quiet public opinion, the government felt that the floods had to be controlled. The centerpiece of flood control was construction of a huge sediment catchment basin in the lower reaches of the Watarase River, just before it joined the Tone River. This meant that Yanaka village, an old and prosperous village, which was a center of the protest movement, would have to be inundated. The catchbasin was completed in 1907, and this, along with tax reductions for affected farmers, succeeded in dividing the locals and defusing political tensions. Thereafter, Ashio died away as a major national issue. Similar problems at other mines, though, were cropping up elsewhere in the country. However, the national government was now prepared to handle them.

Meanwhile, during the heyday of the crisis, Ashio was the site of modern Japan's first clearly documented instance of direct sulfur dioxide and indirect acid deposition damage. Sulfur dioxide and acid deposition damage occurred in the vicinity of the mine's smelter. The main smelter was located in a deep valley surrounded by high mountains. It was constructed in 1884. From the time of its construction the village of Matsuki, which was situated up-valley from the smelter site, began experiencing smoke damage. Matsuki was founded in the Edo period. It was the largest village in remote Ashio township, and was the only large, non-mine-related population center in the vicinity of the smelter. There were forty homes and some 300 people living there in 1892.

By 1893, some fifteen years after mining activities expanded, the forests around the smelter had been totally destroyed. Logging and a major forest fire in 1887 decimated what wasn't damaged by smelter smoke. A prefectural survey conducted in 1887 revealed that 13,000 hectares were completely defoliated; 1,000 hectares had no vegetation growth whatsoever; and 300 hectares were so eroded after death of the vegetation that all topsoil had been washed away, and only bare, exposed rock remained (Shoji and Sugai 1992, 27). In the same year, a Bessemer smelter was installed that decreased smelting time from thirty-two to two days. This multiplied the damage. A 1907 monograph on Ashio's troubles written by the social activist and hard-core opponent of the mine, Arahata Kanson, contains the following passage describing the land around the smelter site:

> "Hard on the heels of the expansion of mining operations, trees were recklessly slashed for timber, and poison smoke [from the refinery] reduced the neighboring forest two *ri* [eight kilometers] on a side to barren wasteland, such that in a flash the once ancient, luxuriant mountain forests of the area became rocky and stripped of green, and there was not a leaf left clinging to obstruct one's view" (Arahata 1963 [originally published in 1907]): p. 9 of the first chapter; author's translation).

According to the figures in this quote, some 6,400 hectares of forest had been rendered "wasteland."

Smoke from the Ashio smelter (and smelters elsewhere in Japan) contained two types of poisons: SO_2 and heavy metals in dust particles.

Ashio ore (and most of Japan's ores) had a sulfur content of 30–40% (Shoji and Sugai 1992, 40). In the refining process sulfur dioxide was produced and emitted from the refinery chimney. In addition, there were also trace amounts of arsenic, cadmium, gold, lead, silver, and zinc in the ore that were also released into the air. Continual exposure to these twin smelter smoke poisons—SO_2 and heavy metals—in conjunction with forest fires and heavy logging, were what eventually killed off the forests in a wide area around the Ashio refinery. Once the forest floor was exposed, heavy rains in the area (more than 2,000 mm/yr) washed away the topsoil together with accumulated toxins.

The introduction of the Bessemer smelter greatly increased the smoke damage to Matsuki village. Crops were damaged, horses and people became ill, silkworms died, and the mulberry trees whose leaves were fed to the silkworms withered. In an 1897 directive issued by a government-initiated Ashio Copper Mine Poisons Investigation Committee most measures were directed at the downstream water-related damage, but a few addressed the smelter operation. One such measure was an order to close a smaller refinery in a neighboring valley and consolidate all smelting operations in one location, the smelter below Matsuki. This, of course, was devastating to the village. Another was an order commanding the construction of a "condensation tower" to reduce the sulfur emissions from the refinery. At great expense a new smokestack was built that attempted to cleanse the smoke of SO_2 with limewater, but it was a dismal failure. It was a highly innovative and admirable attempt to solve the problem by technological means, but nowhere in the world at this time was there an effective method for removing SO_2 from stack emissions; this technological advance lay decades in the future. Because of the failure to control SO_2 emissions, forest damage intensified and the situation at Matsuki became desperate. Villagers petitioned the government for relief to no avail. Eventually Furukawa Ichibei purchased the village lock, stock, and barrel. It was abandoned and demolished in 1901, and thereafter turned into a massive mining waste dumping ground.

Once Matsuki village was eliminated the Ashio smelter had essentially free reign to pillage the local environment.[5] As already mentioned, major prevention measures to control downstream pollution damage were carried out, but little was done to contain damage due to smelting operations. After the attempt at desulfurization this technological avenue was abandoned. Little effort was directed at reforestation or stopping soil erosion. Only in 1987, one hundred years after initial damage began, did the Furukawa Mining Company finally acknowledged its negligence in relation to smelter damage. It formally apologized to residents of the area and prefecture. The company then helped formulate a so-called "Gaia Plan" with local and national government officials to restore the valley to a state of green, which some speculate will take at least a hundred years. The amount of remedial construction work today around the now-defunct smelter is amazing—dams, catch-basins, landslide barriers, terraces of young plantings covering whole mountainsides—and compared to the rich green of the ancient forests in neighboring Nikkō, Ashio is an eye-catching desert yellow. When I first visited Ashio in the spring of 1995 it reminded me not of any mountains I had seen in Japan, but of those in central Asia or the American Southwest.

Besshi Copper Mine (1893–1910)

Similar to Ashio, the Besshi Copper Mine in Ehime Prefecture on the island of Shikoku was the site of major environmental pollution.[6] However, at Besshi, smelter emissions not waterborne pollution was the prime source of damage and motivation for citizen activism. Besshi was owned by the Sumitomo family and is the original source of wealth of today's huge Sumitomo multinational corporation. Copper mining began at Besshi in 1691 and continued for 282 years before it ceased in 1973. As in the case of Ashio, modern mining operations were instituted soon after the Meiji Restoration. A French engineer, Louis Larroque, completely overhauled the old system in 1874.

During much of its early history, local farmers around the Besshi smelter, which was at first located in the mountains at the mine site, experienced crop damage during smelting operations. But the scale of complaints escalated when in 1888 a new western-style smelter was built in Niihama city on the Inland Sea coast a short distance from the mine. From the beginning farmers were strongly opposed to the new smelter. After start-up, the Niihama smelter lived up to the farmers' fears. Significant crop damage soon manifested. Protests to the prefectural government succeeded in persuading the Ministry of Agriculture and Commerce's Osaka Bureau of Mines to confirm the existence of crop damage, but the company denied there was a connection between smelter emissions and crop damage, and refused to consider compensation.

Protest was quieted during the Sino-Japanese War of 1894–1895. After the war, though, smoke pollution worsened, crop damage increased, protests flared, and the Ministry of Agriculture and Commerce ordered the company to bring the problem under control. This order was contained in the previously mentioned 1897 directive, which was aimed at all mines in Japan. The Ministry, through its Osaka Bureau of Mines, ordered Sumitomo to move smelting operations. This was done some seven years later when a new smelter was completed in 1905 on the island of Shisaka about 20 kilometers from Niihama city. But the result was an even wider area of impact. Smoke from the refinery traveled along the coastline causing extensive agricultural damage. Sumitomo, however, maintained its long-standing position of refusing to pay compensation. There were other problems at Besshi. Following on the heels of labor strikes at Ashio, in 1907 some 1,000 miners at Besshi struck, demanding higher wages and improved working conditions. The army was called in and the strike was broken.

With social unrest threatening to get out of hand at Besshi and other copper mines throughout the country, the Ministry of Agriculture and Commerce set up its third Copper Mine Poisons Investigation Committee in 1909. The government deemed it absolutely necessary to maintain harmony between agriculture and industry. The committee sent technicians abroad to seek out the best foreign technology to control copper smelter smoke. They also sent specialists to the various mines around Japan to assess the damage.

In 1910, the Ministry finally pressured Sumitomo into admitting that its smelter operations were responsible for crop damage, and forcefully stepped in to mediate between the farmers and the company. Sumitomo officials and farmer representatives were summoned to the official residence of the Minister of Agriculture and Commerce in Tokyo, and over a twenty-day period negotiations were held. In the end Sumitomo reluctantly agreed to limit copper production, pay compensation to area farmers, and make adjustments to smelter operations in order to ameliorate the smoke damage. These were groundbreaking measures. The agreement marked the end of the most intense phase of environmental degradation and local farmer protest at Besshi.

Damage compensation negotiations were periodically renewed between farmers and Sumitomo from this time to 1925 (they were renewed a total of six times).[7] The high cost of payments caused Sumitomo to diligently seek technological means of controlling sulfur emissions. Partial control was achieved in 1929, when a new 48-meter stack was built and a reasonably effective system for removing sulfur was discovered. However, the problem was not fully corrected until 1939. Compensation payments were terminated in this year. A total of forty-five years elapsed between the first major complaints and final solution. A state-of-the-art refinery was constructed in 1971; however, the copper mine was closed in 1973 and the new refinery uses imported copper ore.

Although the problem at the Besshi mine did not reach the epic scale of suffering at Ashio, nevertheless the same type of political pattern prevailed. The three main actors were the government (principally the Ministry of Agriculture and Commerce), industry (Sumitomo), and local residents (farmers whose crops were damaged by smelter smoke). The stages of buildup of the problem were similar to Ashio. Farmers first registered complaints with the offending company, Sumitomo. The company totally ignored their pleas. The farmers then took their grievances to the local government, and then to the Ministry of Agriculture and Commerce. The resolution of the Besshi problem was achieved with more alacrity than at Ashio, in part because of lessons learned at Ashio. The Ministry of Agriculture and Commerce sternly coordinated a face-to-face negotiated settlement between Sumitomo and the farmers in 1910, and this basically ended the problem.

Scientists and scientific knowledge do not seem to have played a major role in the Besshi problem. Although the third Copper Mine Poisons Investigation Committee did send a team to investigate in 1909, its purpose was merely to survey the extent of damage, not inquire into the physical-chemical dynamics of the environmental problem. Technology forcing, however, played a role in solving the problem. After the 1910 negotiated settlement, the high compensation payments required of

Sumitomo acted as a technology-forcing mechanism and methods were urgently sought for cleansing the smoke of its sulfur. Until a technological solution was found, the government imposed on Sumitomo a policy of limiting copper production and seasonally adjusting smelter operations to minimize damage at crucial points in the rice cultivation cycle, especially in the spring after planting of rice seedlings. These were innovative policy measures at the time.

Hitachi Copper Mine (1907–1915)

The case of the Hitachi Copper Mine is completely opposite to that of the Ashio and Besshi mines, and is unprecedented in early modern Japanese history as an example of an enlightened approach to environmental problems.[8] The mine is located on the Pacific coast in Ibaraki Prefecture about 120 kilometers northeast of Tokyo. In 1905 Kuhara Fusanosuke, the new owner of the Hitachi Copper Mine, set about modernizing and greatly expanding the mine's operations. Although it had been discovered in 1591, Hitachi was basically nonproductive when Kuhara bought it. However, within three years it became the fourth largest mine in Japan after Ashio, Besshi, and Kosaka.

In 1907 soon after completion of an electric power plant and a new blast furnace, smoke from a 28-meter-high smelter smokestack (called the "eight-sided stack" because of its octagonal shape) resulted in the first serious damage to crops in the vicinity. Since operations were just beginning to expand, Kuhara knew smoke pollution would become a major company problem, and, unlike at Ashio and Besshi, he chose to confront it immediately and directly. He soon initiated negotiations with area residents on how to deal with it. From the very beginning Kuhara undertook two prevention measures. First, he bought up land where pollution was expected, and second he agreed to pay compensation to farmers for any damage incurred. To deal with the compensation issue, representatives from surrounding villages met with company officials and a compensation contract was hammered out and signed with each village.[9]

However, emissions were greater than expected and extensive crop and forest damage resulted. Thus, in 1911 a second "stack" was built. Rather than a standing structure, it was a long chimney, almost 2 kilometers, following the lay of the land up a mountainside. Along its course were dozens of holes for smoke to escape. When viewed from a distance it resembled a giant centipede as smoke emanated from the many holes, hence its nickname: the "centipede chimney." No records remain explaining why this design was attempted but it seems to have been the thought of the original architects that smoke emerging from many holes would dilute and disperse more readily than from a single stack port. However, the heavier-than-air gases, rather than blowing away in the wind, settled in the valleys and caused even greater damage.

By 1912 crop and forest damage extended to over twenty surrounding towns and villages. The village of Irishima was especially hard-hit, and ironically, of all the crops suffering damage it was tobacco plants,[10] for which the area was noted, which faired the worst. Kuhara's compensation payments grew by leaps and bounds. In the period between 1905 and 1912 copper production jumped from about 200 to 8,000 tonnes, and smoke damage and compensation payments grew roughly in the same proportion. In 1912, largely in the wake of the problems at Besshi, there arrived a government order from the Ministry of Agriculture and Commerce to reduce smoke damage to agricultural crops. The order, based upon the recommendations of the third Copper Mine Poisons Investigation Committee and known as the "smoke gas concentration limitation order (*hien gasu nōdo seigen meirei*)," required Hitachi, as well as Ashio, Besshi, Kosaka, and other mines, to limit the concentration of SO_2 in smelter exit gases, especially during certain seasons so as to minimize crop damage. The technical solution that the committee devised to dilute SO_2 concentrations was construction of short, fat stacks!

Because of Hitachi's experience with the unsuccessful centipede chimney, company engineers seriously doubted the effectiveness of the government-ordered stack but were forced to comply anyway. Thus, a new low, thick stack was built in 1913 with a height of 38 meters and a diameter about half the height, 18 meters. This was Hitachi's third stack. The government also required the use of powerful fans inside the stack to dilute stack gases with outside air, but because these fans demanded more power than the electricity-short company could afford, it obtained permission to eliminate the fans if they could meet the dilution requirements without them. But the stack was a disaster. Instead of dispersing in the air, SO_2 formed a thick layer near the ground after cooling. This only magnified ground level concentrations. In addition, sulfuric smoke not only exited the stack but also, because of its huge girth, was often forced back into the smelter building under certain meteorological conditions, causing illness among the workers. The stack had many nicknames. Perhaps the most apt was "stupid stack (*ahō entotsu*)." Others included "tank chimney," "*daruma* chimney" (after the short, fat *daruma* dolls of Japan), and "government-order chimney." Hitachi built only one such stack though government orders called for three. This type of short, fat stack was also forced upon other mines with similar results, and was soon abandoned. In view of the failure of the government-ordered stack, Hitachi took to limiting copper production depending on season and weather conditions. Yet because of expanding operations smoke damage only worsened.

No one is sure who came up with the idea, but given the experience with the first three stacks (the eight-sided stack, the centipede chimney, and the stupid stack), company officials began debating the merits of building a very tall stack. In 1914 company scientists, as a result of meteorological observations (to be explained later), discovered upper level air currents that blew with more directional consistency than surface winds and that often blew out to sea. Their recommendation was that a tall stack would be far more effective in dispersing pollutants because it would be like "placing the stack in the middle of a river current." Contrary to Ministry of Agriculture and Commerce orders, Kuhara resolved to construct a stack taller than anyone else had ever built. He reasoned that even if it was a failure it would be worth the effort just to learn the technology. The result was a 156-meter-high stack with a 27-meter-diameter base set on a 4-meter-thick foundation. The inner diameter at exit was 7.8 meters. The stack was constructed in nine months and was designed to withstand earthquakes, high winds, and lightening. Completed in December 1914 and fired up for the first time in January of the next year, it was the world's tallest smokestack at the time.[11]

Upon completion of the "Big Stack (*Dai Entotsu*)," as it was called, SO_2 damage and resulting compensation payments instantly plummeted. Only a few months after the Big Stack came on-line, the leaders of the farmers' antipollution group declared the era of smoke damage over.

During the decades after 1915 the "Big Lonely One," as one song affectionately named the stack, became a symbol of the city of Hitachi and rendered faithful service to a series of increasingly nonpolluting smelters until its upper two-thirds suddenly toppled in 1993.[12]

The Hitachi story is not yet complete. Besides the technical accomplishments of the Big Stack and its success in eliminating local SO_2 damage, there were a series of other extraordinary scientific and environmental accomplishments. To pick up these threads we need to return to 1909, when meteorological observations began at Hitachi. What started as a single meteorological station eventually grew into a private network of seven stations covering a several hundred square kilometer area around the mining complex. It was the first meteorological network in Japan. One of its main functions was to gather data on the relation between the "smoke path" and meteorological conditions to forecast those times when smelter operation should be cut back. (Even after the Big Stack was completed, the network was used to track the occasional touchdowns of the plume and notify officials in charge of smelter operations to reduce emissions.) The most interesting set of meteorological observations at Hitachi, though, were not those of the network, but measurements of high altitude wind speed and wind direction.

The impetus for the first measurements seems to have emerged in conjunction with talk about building a tall stack. The data were to be used to determine exactly how tall the stack should be built. However, at the time no one had any idea how to make high altitude wind speed and wind direction measurements. This had never been done in Japan. One of Hitachi's scientists, Unno Kiyoshi, hit upon a method when he saw a small dirigible brought up from Tokyo for a summer festival at Hitachi—he would use the dirigible to loft up instruments to take the measurements! He then studied dirigible use with the army, which had first employed them during the Russo-Japanese War of 1904–1905. Unno purchased one from the army, and began taking measurements of air currents. It was he who discovered the strong upper air currents, which often blew seaward, and this apparently led to the final decision to build a superstack. A special station and dirigible hanger was built for the high altitude observations in 1915. They continued until 1919 after which they were terminated.

We still have not completed the Hitachi story. It was also in 1909, the same year the meteorological network was established, that under the leadership of Kaburagi Tokuji an agricultural station was constructed for testing the effects of SO_2 on plants.[13] Kaburagi conducted a wide range of indoor and outdoor experiments. For the indoor experiments he used a highly sophisticated little greenhouse of his own creation for artificially exposing plants to SO_2. In the greenhouse, SO_2 was produced in the anterior third of the structure and blown into the front two-thirds, which was a glassed-in room where potted plants could be set. In this way he was able to test a wide variety of tree seedlings and crops for their tolerance to SO_2 exposure. He ranked crops into one of twelve classes according to their resistance to SO_2 damage. His experiments revealed that the most SO_2-tolerant crops were corn, millet, chestnut, and sugarcane, and that the least tolerant were buckwheat and tobacco. He similarly classified evergreen and deciduous tree species. On the basis of his extensive experiments he freely and enthusiastically offered advice to area farmers, and as a result became highly trusted and respected by the local community.

Kaburagi's accumulated knowledge of SO_2-tolerant and acid-tolerant plant species proved invaluable even after Hitachi's fourth stack, the Big Stack, was completed. The first three stacks had turned the area around the smelter complex into a wasteland. Under Kaburagi's guidance a massive reforestation program was embarked upon.[14] The company planted some five million tree seedlings (especially Oshima cherry, black pine, and Japanese cedar), and distributed free to area residents another five million. In the worst impacted areas near the smelter complex where the soils had become acidified, lime was mixed into the soil and seedlings and grasses were planted. Restoration and reforestation work continued for the next fifty years. In stark contrast to Ashio, the visitor today to the former site of the Hitachi Copper Mine would never know that the area had once been rendered a wasteland, so complete is the rehabilitation.

The intimate connection between science and company policy at the Hitachi Copper Mine was unprecedented. The measures taken by Hitachi to counteract severe smelter emission damage included: damage surveys, generous compensation terms, production cutbacks (seasonal and weather-dependent), weather observations (seven-station meteorological network and high altitude dirigible measurements), crop and tree research (conducted outdoors and also indoors in a sophisticated little greenhouse), agricultural extension-type outreach to area farmers, and massive reforestation and erosion prevention projects. Given the creativity exhibited at Hitachi it is perhaps not surprising the corporate legacy that Kuhara bequeathed to the future. After the Hitachi Mine was taken over by his brother-in-law in 1928, it was reorganized as Nippon Sangyō Company—better known today as Nissan. Also, today's multinational giant, Hitachi Electric, was borne in the electrical equipment and maintenance shop at Hitachi. In addition, one of Japan's leading oil companies, Jomo, is a direct descendent of Kuhara's later oil ventures. Present-day corporations would do well to study the environmental lessons of the Hitachi Copper Mine.

Early Precipitation Chemistry

1883: First Precipitation Measurements

According to Kurashige (1934), a German professor, Osker Kellner, conducted the first measurements of precipitation chemistry in Japan between 1883–1885. Symbolically, we could call this the first instance of localization to Japan of universal acid deposition-related knowledge, knowledge carried from its birthplace in Europe. Kellner was temporarily hired by the Agricultural University in Tokyo (the forerunner to today's College of Agriculture at Tokyo University). The purpose of his analyses was to determine nutrient inputs from rainfall to agricultural crops. Kellner measured ammonia and nitrate concentrations in rainwater, and published the results in two locations (Kellner 1887, 1894: the content of these two publications is identical).

Kellner came to Japan in 1881, and in addition to teaching at the university he conducted experiments for the Ministry of Agriculture and Commerce. Four of his students—Suzuki Zenjirō, Sawano Jun, Makino Kenzō, and Yoshii Toyozō—aided in the precipitation chemistry analyses. All of these students later became prominent agricultural chemists in Japan. Yoshii translated Kellner's papers referred to previously. Kellner, today regarded as one of the founding fathers of agricultural science in Japan, returned to Germany in 1892 where he continued an illustrious career.

1912: First Urban Precipitation Chemistry Measurements

The earliest record I found of precipitation chemistry measurements in an urban area date from 1912 in Osaka (the measurements are briefly summarized in Akimoto (1929) some seventeen years later). Given Osaka's "king of air pollution" status in Japan throughout the prewar period it is not surprising that the first urban precipitation chemistry measurements seem to have been conducted here. The Osaka Public Health Laboratory (*Osaka Shiritsu Eisei Shikenjo*) first monitored air pollution (soot and smoke) around this time, and analysis of rainwater for pollutants seems to have been a part of this program. Hirayama, the researcher who performed the measurements, analyzed precipitation for acidity, chloride, ammonia, and sulfate content, and the results were published in the laboratory's bulletin. He found, for instance, high sulfate levels in the city; there is no indication that these measurements were continued after 1912.

1913: Nishi-ga-Hara Record of Precipitation Chemistry

Kellner's work on agriculture-related precipitation chemistry continued at four agricultural experiment stations run by the Ministry of Agriculture and Commerce—Nishi-ga-Hara outside Tokyo, Kumamoto in Kyushu in southern Japan, Kashiwabara near Nagoya in central Japan, and Omagari in northern Japan. The person in charge of these measurements was Kawashima Rokurō, a student of Kellner's original disciples. The Nishi-ga-Hara data are the longest continuous record of rainfall chemistry (sulfate, ammonia, chloride, nitrate, and nitrite) in early modern Japan. They began in 1913 and continued up to 1934 (Kurashige 1934). The Nishi-ga-Hara data were from bulk samples that captured both dry and wet deposited substances. Although their purpose, like the measurements of Kellner, was to assess nutrient inputs from rainfall, a couple of interesting trends related to Tokyo air pollution were observed. First, the yearly averages in sulfate seemed to reflect ups and downs of industrial activity in the Tokyo area, and second, the sulfate concentrations in fog and drizzle were remarkably higher than for ordinary rain. Fujita (1993a, 74) compared the Nishi-ga-Hara nitrate data to nitrate concentrations in the same location in 1975 (the peak year for NO_x emissions in Japan) and estimates that the concentrations increased threefold in sixty years.

Analysis of Period 1

The Meiji Restoration of 1868 marked the end of Japan as a "sustainable society." During the early part of the ensuing Meiji period (1868–1912), a wave of exuberant reform—*bunmei kaika* (enlightened civilization)—swept the nation. The old, sustainable Japan was seen as weak, backward, and embarrassing. The strength of the new Japan seemed vindicated with victory in the Sino-Japanese War in 1895. However, soon after this victory less desirable elements of the reforms came to light. Pollution incidents throughout the country called into question the sustainability of Japan's new found industrial strength.

The first and most explosive environmental problems centered on air and water pollution at copper mines.[15] Copper during the Meiji and early Taisho (1912–1926) periods was the third largest foreign exchange earner behind silk and tea, both of which were basically nonpolluting industries. Major incidents of human and animal health impacts, and agricultural crop and forest damage, occurred due to sulfur dioxide emissions from copper smelters.[16] Smelter emission damage at the Ashio, Besshi, and Hitachi copper mines was described above. These were not the only cases, but they were the most prominent. All three mines experienced significant damage, which, except at Hitachi, spawned large-scale and often violent protest movements. The protest movements eventually sparked a state of crisis in the early years of the twentieth century. Because of the importance of copper to Japan's foreign exchange earnings and because of the government's desire for popular support in its drive to industrialize, militarize (and imperialize),[17] strenuous efforts were made at the highest levels of government and industry to harmonize the industrial need for copper and the political need for citizen support. One result was the birth of acid deposition science in Japan. What was the structure of early acid deposition science? Did an acid deposition problem-framework emerge? Did an acid deposition expert community coalesce? How did the science mix with the politics? And how did scientists and policymakers together address the question of sustainability? To answer these questions, we must first answer the question: how and why did science itself take hold in Meiji Japan? The origin of acid deposition science falls within the larger context of the general development of science in early modern Japan.

Meiji Science

During the Meiji period modern science was imported wholesale from the West into a society that had little experience with the ways of thinking embodied in this practice. Science was only one of a myriad of new ways of thinking thrust upon Japan. The two dominant modes for import of science were (1) to invite foreign teachers to Japan and (2) to send students abroad for study. Between 1868 and 1900, the Japanese government employed over 8,000 foreigners in teaching and other capacities (Bartholomew 1989, 64). Foreign professors teaching at Tokyo University were the elite among this group (Kellner is one example).[18] They wielded great influence in shaping the early course of Japanese science. The extent, and eventual decline, of this influence is illustrated in figure 4.1. As depicted in the figure, between 1870–1875, Westerners wrote all of the articles in the newly established natural science journals in Japan, but thereafter contributions steadily declined until by 1887 only about 10% came from Westerners.

Foreign professors were the primary vehicles for initial transmission of scientific knowledge to Japan (and localizing scientific knowledge to Japan), but they were expensive vehicles. In the 1870s and 1880s their salaries were one-third or more of Tokyo University's budget (Bartholomew 1989, 64). In part to replace costly foreign scientists, the Japanese government in the 1880s began emphasizing overseas study, as table 4.1 illustrates (note the huge increase in study abroad during the 1883–1914 period). Between 1869 and 1914, 66% of all student-years were spent in Germany, 11% in Great Britain, 11% in the United States, 7% in France, and 5% in other countries (Bartholomew 1989, 71). This pattern held for almost all fields of science.

One characteristic of Meiji science that shaped acid deposition science was its overwhelming emphasis on practical knowledge. Medicine,

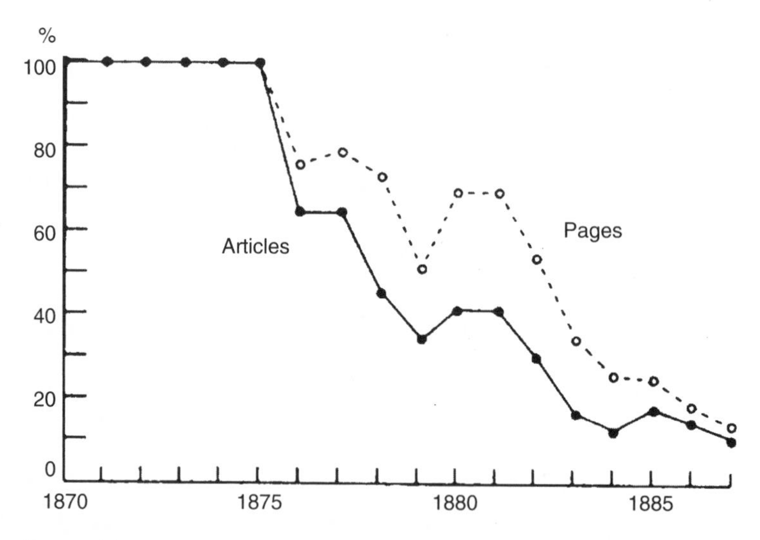

Figure 4.1
Western contribution to academic articles on the natural sciences (1870–1887). Reprinted from Masao Watanabe, *Science and Cultural Exchange in Modern History: Japan and the West* (Tokyo: Hokusen-Sha, 1997), p. 211, diagram III. Used with permission of the author.

Table 4.1
Overseas study in early modern Japan (doctorate recipients)

	Student-Years Abroad
1867–1877	138
1878–1882	163
1883–1914	2,842
1915–1920	418
Total	3,561

Note: Total student-years = number of students times number of years each student studied abroad.
Source: Bartholomew (1989, 69).

Table 4.2
Number of doctorates in technical fields (1888–1920)

Fields	Number
Agriculture	108
Biology	56
Chemistry	27
Engineering	366
Forestry	30
Geology	18
Mathematics	22
Medicine	656
Physics	54
Veterinary medicine	23

Source: Bartholomew (1989, 52).

engineering, and agriculture were the favored applied sciences (see table 4.2). In contrast to the historical development of science in the West, Japanese science was essentially born from these disciplines, not physics, mathematics, and astronomy as in the West. Indeed, in early modern Japan there was little differentiation between science and its applied cousin—technology. "The conceptual schemes of Western science . . . were taken as true because they came from the West, and Meiji commentators gave little attention to differences between science and technology, regarding them as the same thing" (Bartholomew 1989, 4). To the Japanese the goal of science was to practically benefit society. Thus, it should be no great surprise that acid deposition science initially emerged from agricultural science, one of the main fields of 'practical' inquiry in Japan.

The emphasis on applied knowledge meant that empirical observation was valued over theory building. In the case of acid deposition, numerous and varied observations were made regarding the impact of copper smelter smoke on surrounding people and landscapes, for instance, but no cause-and-effect theory of the emission, atmospheric transport, and deposition of pollutants emerged. In the West (as will be related shortly) such a theory for industrial emissions originated for the first time in England around 1850.

A second characteristic of Meiji science that shaped acid deposition science was the commitment of both scientists and government officials to the same basic ideals related to establishing a new and strong Japan. The historian of science, Yoshida Mitsukuni, wrote: "We should remember that among the scientists [of Meiji Japan] was the desire to participate in building a new era" (as quoted in Bartholomew 1989, 4). As with government bureaucrats, scientists were elites primarily recruited from the samurai class. They attended the top universities, of which Tokyo University was the pinnacle.[19] More so than later periods, scientists during the Meiji period were coequals with bureaucrats and acted as intimate advisors on technical issues. To the degree that I was able to ascertain, scientists and engineers investigating acid deposition-related problems were indeed treated as equals. Their lack of political influence was less the result of differences in status than the lack of a compelling theory for understand acid deposition phenomena (or, in the language of this book, the inability to construct a problem-framework that could serve as a template for scientific understanding and political action).

A third characteristic of science in early modern Japan that shaped acid deposition science was its amorphous and uncertain nature. Science was imported "raw" into a scientifically virgin land. There was great uncertainty about what to study and how to study it. The task of Japan's first generation of scientists between 1868 and 1920 was not only to digest Western scientific knowledge itself, but also to apply it to the specific circumstances of Japan. The process of applying it in many cases was sporadic. Thus, there are many examples of scientists conducting a particular scientific analysis only once. Instead of the work continuing, it seems to disappear from the record (the work of Hitachi's Kaburagi and Osaka Public Health Laboratory's Hirayama are examples of this). In general, Japan was experimenting with science.

A fourth and final characteristic of Meiji science that shaped acid deposition science was the highly contested and largely undefined role of researchers within society. This role did not begin sorting itself out until about 1920. "Universities were founded, laboratories built, and academic societies established; but the researcher role remained vaguely defined, inchoate, and surrounded by hostile forces" (Bartholomew 1989, 82–83). Community formation in science (encompassing the processes of recruitment, training, and socialization of scientists) took root slowly and did not reach a turning point until World War I.

Prior to the war, Japan depended on Germany for most of its flow of ideas. The war blocked the flow and forced a transformation. Instead of relying on outside ideas, Japan had to generate its own. The government expanded domestic funding for research in both basic and applied sciences. One effect was that instead of solely working on matters of immediate practical importance, scientists were freer to investigate a wider range of phenomena. This helped define what it meant to be a scientist in Japan, and assisted in the process of expert community formation. The new role included openness to scientific criticism, avoidance of favoritism, and recognition of the public nature of science. In fact, the changing role of scientists evident around 1920 coincides with the historical transition from the first to the second acid deposition period. "Japan was not a center [of science] by 1921, but it had reached the point of self-sustaining growth and stood on the threshold of major contributions" (Bartholomew 1989, 8). One of these contributions was the flowering of precipitation chemistry to be described in the next chapter.

In summary, between 1868 and 1920 the cognitive structure and practice known as science was transmitted to Japan from the West (primarily from Germany, Great Britain, the United States, and France), and its local adaptation to Japan began. The main points of entry were the elite educational institutions in Tokyo, especially Tokyo University. These institutions in turn exported science to the provinces. Upon arrival in Japan, science was filtered through Japan's historical and cultural context. Imprints from this context on Meiji science include the previously described emphasis on practical versus theoretical knowledge, the blurring of the line between researcher and bureaucrat, and the confusion over the role of scientists in society. "Japanified" or Meiji science provided the backdrop against which acid deposition science emerged.

What kind of acid deposition science emerged in Meiji Japan? To understand its character we need to first examine the import (or lack thereof) of acid deposition knowledge from the West and its adaptation (or lack thereof) in Japan. Thus, in the following section, we turn to the fountainhead of Japan's acid deposition-related ideas—Europe and North America—and survey the state of science in these two regions. This will give us a notion of what was available for potential transmission to Japan.

Acid Deposition in Europe and North America

Acid deposition-related work in Europe was far in advance of Japan.[20] Robert Angus Smith (1817–1884)—working in the heartland of the industrial revolution, England—formulated in the latter half of the nineteenth century a basic outline of the physical and chemical cause-and-effect relationships that resulted in acid deposition problems. In 1852 he published a detailed report on air quality in Manchester titled *On the Air and Rain of Manchester*. Smith was appointed the first Chief Alkali Inspector of the United Kingdom in 1863, and thereafter had the opportunity to observe the connection between industry and pollution even more closely.[21] Twenty years after publication of the Manchester air quality report, and based on detailed observations in England, Scotland, Ireland, Belgium, and Germany, Smith published in 1872 what has been called "an extraordinary book" (Cowling 1982, 111A)—*Air and Rain: The Beginning of a Chemical Climatology*. In this work he coined the term "acid rain" and deduced some of the principle ideas that underlie the study of precipitation chemistry and its relation to acidification phenomena.[22] In particular, he concluded that the chemical composition of precipitation was influenced by such factors as coal combustion and decomposition of organic matter, wind trajectories, distance from the coastline, and amount and frequency of rain or snowfall. He proposed a detailed set of procedures for the proper collection and analysis of precipitation. And he noted the damage acid rain caused to plants and materials. In other words, Robert Angus Smith pieced together, a hundred years before acid deposition was recognized as a significant environmental problem, the basic scientific elements of the phenomenon (emission, atmospheric processes, and environmental and material impacts). But Smith was so far in advance of his time that, unfortunately, his work lay in obscurity until almost a century later.

After publication of Smith's books, and during the decades leading up to World War II, scientific knowledge on acidification processes, and on the link between fossil fuel combustion, primary metal smelting, and acid deposition, grew slowly as a result of diverse research in Europe. The knowledge gained was for the most part disperse and unconnected. No attempt is made here at a complete record of this research. To give a few examples of the type of knowledge being accumulated, though, a Belgian commission conducted extensive studies in 1854–1855 on damage to plants from acidic emissions from chemical industries; the famous Rothamsted Experiment Station in England conducted research on the relation between crop growth and nutrients in the atmosphere (via precipitation chemistry analyses) from 1855 to 1916; in 1909 Sørensen of Denmark invented the pH scale to describe the acidity of aqueous solutions.

In 1911 C. Crowther and A. G. Ruston in England studied the effects of acid inputs to soils around the city of Leeds based on measurements during 1907–1908 (they attributed the strongly acidic precipitation to coal combustion and showed that acidity levels decreased away from the center of Leeds). In 1919 Rusnov of Austria reported on the acidification of forest soils due to SO_2 emissions; and in 1921 Dahl of Norway recognized a relation between surface water acidity and changes in trout production. Perhaps most significantly, the field of precipitation chemistry as it relates to agriculture was well established in Europe and North America by the late 1800s and early 1900s. It makes sense that Japan first imported precipitation chemistry via agricultural science. Thus, agriculture-related precipitation chemistry was highly important in establishing acid deposition science in both Japan and the West. In fact, in Europe it eventually led to identification of the world's first large-scale acid rain problem in the 1960s.

In North America (Canada and the United States), science in general, and acid deposition-related science in particular, lagged behind that of Europe. In North America, as in Japan and almost at the same time, the first major instances of acid deposition damage were due to mining operations. In Canada the mammoth mining complex at Sudbury, Ontario induced the first notable mining pollution problems in the late 1800s.[23] The Sudbury copper and nickel mining complex was opened in 1886 and contains a magnificent copper deposit and the world's richest nickel deposit. By 1916 the Sudbury mine was pumping an estimated 600,000 tonnes of SO_2 into the air.

Just as in Japan, irate farmers experiencing health problems and crop damage attempted to force improvements in the situation. And just as in Japan, the Canadian government response was biased toward the powerful mining interests. Japan in the early twentieth century witnessed the notorious destruction of Yanaka village due to mining activities at Ashio; Canada in the early twentieth century witnessed the declaration by the government of twelve townships around Sudbury unfit for cultivation and habitation. However, given its massive size Sudbury proved to be an ideal field laboratory for the study of the negative impacts of acidic emissions. The first studies on plants and soils in the area were carried out in the 1940s.[24] Similar studies were conducted at the site of the Trail Smelter in British Columbia in the 1930s.[25] Other mines in North America (especially those in the United States) were generally spared the controversy of Sudbury because of their remote location away from populated areas.

In summary, a fairly deep pool of Western acid deposition knowledge existed during Japan's first acid deposition period. Europe gave birth to acid deposition science and was the leader. North America followed. In England Robert Angus Smith actually succeeded in constructing a crude problem-framework. However, his framework did not take hold, and acid deposition remained indistinctly folded into the broader problem of air pollution caused by industrialization. Thus, we can conclude that at the time that Japan was absorbing Western science in general, it also had a rich but rag-tag collection of Western acid deposition-related knowledge to potentially draw upon. How much and how did it draw on this knowledge?

Beginnings of Acid Deposition Science

Between 1868 and 1920 Japan developed what is best called *proto-acid deposition science*. Absorption of acid deposition-related ideas from the West was sporadic, uncoordinated, and minimal; it derived initially from foreign teachers and later from foreign journals, and it took place primarily at Tokyo educational institutions. No coherent problem-framework for the smelter smoke problem existed in the West for Japan to import.[26] Japanese scientists tackled bits and pieces of the problem, but were never able to put the fragments together. A vast amount of data (on emissions, meteorological conditions, and human and plant impacts) was collected but it was never synthesized. Activities at Hitachi came the closest to achieving a synthesis, but even here, once the worst of the damage was alleviated (by building the world's tallest stack), the research ceased. Thus, no coherent understanding of acid deposition-related phenomena resulted in Japan—only fragments of a problem-framework and an expert community were visible.

Even though pollution from the city of Tokyo was measured at Nishi-ga-Hara using the methods of precipitation chemistry, I was unable to discover any record of precipitation chemistry measurements carried out at any of the copper mines. Thus, the connection between sulfur dioxide emissions from smelters and acidified rainwater seems not to have been made, although it was, of course, known that such emissions acidified soils, as attested to by the fact that Hitachi's Kaburagi sought out acid-tolerant plant species and added lime to soils in Hitachi's reforestation projects.

The acid deposition-related scientific knowledge base for copper smelters was fairly simple. It rested on the fact that the predominant damage-causing agent in emissions from smelters was sulfur dioxide. That sulfur dioxide was emitted was never a major source of controversy (this could readily be determined by the chemistry of the day), but that sulfur dioxide caused damage was indeed a major source of controversy. As we saw in the Besshi case, Sumitomo vehemently denied that sulfur dioxide caused crop damage and therefore refused to pay compensation to area farmers. The fact of a relationship between sulfur dioxide emissions and environmental damage was validated not by science as much as by the bitter personal experience of residents in the vicinity of smelters. However, on-site surveys of the extent of damage and correlations between weather conditions and damage gave it a firmer "scientific" foundation. The only instance of actual experiments on sulfur dioxide damage to vegetation of which I am aware are those conducted by Kaburagi at the Hitachi Copper Mine.

In the end, after the advent of multiple cases of plant, animal, and human injury, the correlation between sulfur dioxide emissions and environmental damage became an accepted fact in Japan around the turn of the century. However, it did not become fact because scientists constructed a coherent problem-framework and created bridging objects to facilitate communication between scientists, bureaucrats, company managers, and the general public; it did so because of the obvious extent, intensity, and nature of the problem. The science of the day was not up to ferreting out the detailed mechanisms by which damage occurred, and thus was not up to spelling out an authoritative solution path.

Beginnings of Acid Deposition Policy

The government framed the copper smelter smoke issue as one of how to control the smoke and minimize damage such that copper production could continue relatively unimpeded. To assist in formulating policies, the government formed (between 1897 and 1909) three copper mine poisons investigation committees consisting of scientists, engineers, and government officials. In addition to government efforts, companies attempted to devise solutions on their own. In general, the flow of information (scientific and otherwise) between the government and mining companies, and between the mining companies themselves, seems to have been fairly extensive. But the information flow between the government and mining companies, and local residents and the general public, was more restricted. The following is a list of the major successful and unsuccessful policies and countermeasures that attempted to deal with the smelter smoke problem.

• *Desulfurization Technology* Desulfurization technologies were tried, but none were successful. As early as 1897 an unsuccessful attempt was made at Ashio to treat stack emissions with limewater. In 1911 an attempt was made at Hitachi to precipitate sulfuric acid from stack emissions. The government even sent specialists abroad to learn of the latest foreign emission control technologies. Unfortunately though, nowhere in the world did there exist at this time a proven desulfurization method.

• *Stack Technology* Numerous stack technologies were experimented with. The first stacks simply exhausted smelter emissions with no thought given to damage. The third Copper Mine Poisons Investigation Committee came up with the idea of short, fat stacks to dilute sulfur dioxide emissions. Such stacks were installed at several mines, but the results were disastrous. Hitachi was the first to hit on the tall stack solution and built the world's highest stack in 1915.

• *Compensation Contracts* Most mines voluntarily or involuntarily signed compensation contracts with local residents. This became the primary vehicle by which victims of smelter damage were assuaged of their suffering and by which public protest was defused. Financial compensation schemes were to become a hallmark of the Japanese approach to environmental pollution. Such schemes implicitly sanctioned continued pollution and deflected policy measures away from ecological impacts. Japan today possesses the most advanced and sophisticated compensation legislation in the world.

• *Damage Surveys* Damage surveys became more or less routine in the wake of compensation payment contracts in order to determine the range and extent of damage. Some were carried out by the Ministry of Agriculture and Commerce, and others by universities, local governments, and mining companies.

• *Land Purchases of Damaged Areas* Some mines purchased damaged land outright, as in the case of Ashio's purchase of Matsuki village. Land that would potentially be damaged was also purchased, as was the case in Hitachi's purchase of land around its smelter complex.

• *Production Cutbacks* A schedule of seasonal and weather-dependent copper production cutbacks was practiced at most mines. The first negotiated cutbacks were worked out in government-sponsored negotiations between Besshi and local farmers in 1910. This became common practice thereafter. For instance, the government ordered Hitachi to initiate cutbacks in 1912. On occasion, smelter operation was stopped altogether for extended periods.

• *Weather Observations* In part to determine when damage-exacerbating weather conditions threatened, meteorological observations became common at several of the mines. The earliest weather station at a mine was set up at Ashio in the late 1800s. The most sophisticated meteorological network was Hitachi's seven-station network. Using a dirigible, Hitachi also conducted Japan's first high altitude meteorological measurements.

• *Crop and Tree Experimental Research* Hitachi was the only mine to conduct extensive experiments to determine the tolerance of plant species to sulfur dioxide. However, at least some of what was learned at Hitachi seems to have been passed on to other mines. The advice of Hitachi's Kaburagi was sought by other mining companies. Hitachi also seems to be the only mine that practiced an agricultural outreach program to help area farmers deal with smelter emission damage.

- *Reforestation and Erosion Prevention Projects* Hitachi also seems to be the only mine that carried out an extensive program of reforestation and soil erosion prevention projects.

What was the role of scientists in formulating smelter smoke policies? Figure 4.2 illustrates the science-policy interface on the copper smelter smoke problem. The figure does *not* represent the chronological sequence of activities directed toward solving the problem; instead, it represents the major components of the process. In reality, especially in the case of Ashio, once copper mining issues became front-page news, all components of the policy process were in simultaneous motion. Science, as the figure indicates, thinly influenced all components to one degree or another. Science was not needed, however, to determine that a problem existed (i.e., to determine that smelter smoke caused local damage). The problem was so obvious that a scientific cause-and-effect analysis was not critical to initiate policy action. The most influential experts were those who served on the three investigation committees.[27] In addition, other agricultural specialists, doctors, and engineers contributed to the policy deliberations. Still other experts indirectly influenced policy via their connections with the mining companies or local governments.

Scientists and their scientific knowledge seemed to have been tightly bundled in relation to the policy process. In other words, the most prominent scientists brought pretty much the full weight of available scientific knowledge directly to government officials, primarily those in the Ministry of Agriculture and Commerce. There were few if any mediators of scientific knowledge (knowledge brokers) situated between scientists and bureaucrats to transmit and interpret knowledge for policymakers.

The loose community of scientists and engineers who tackled the "sulfur dioxide damage due to copper smelter emissions" problem in Japan around the turn of the century were able to construct a rudimentary scientific foundation that informed a crude set of government and mining company policies. Upon this foundation a coarse "smelter smoke problem-framework" emerged. The roughness of the framework dampened the ability of scientists to form conceptual bridging objects to tie the multiple social worlds working on the problem. Some bridging objects were created; for instance, the tall stack solution suggested by the dirigible experiments at Hitachi, the usefulness of meteorological

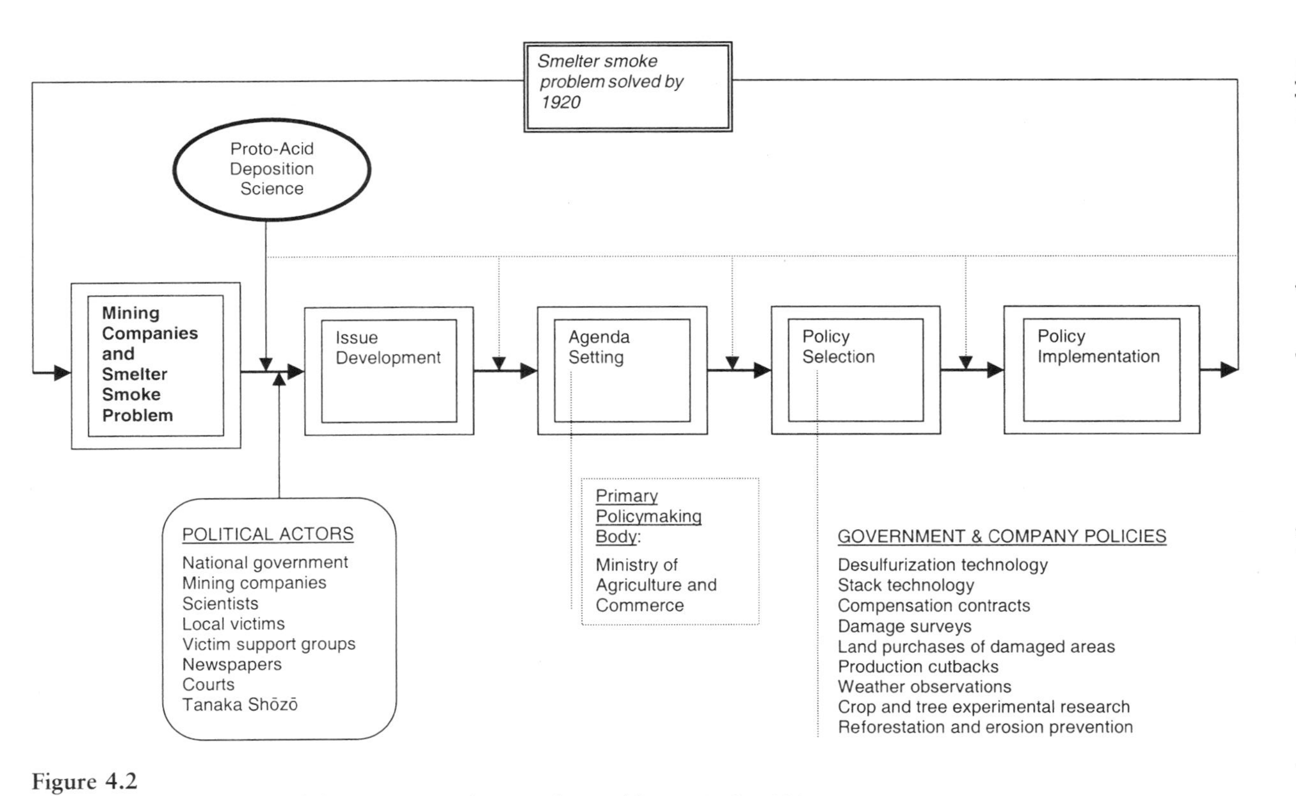

Figure 4.2
Science-policy interface of the copper smelter smoke problem (1868–1920)

observations to forewarn of plume touchdowns, and the planting of acid tolerant crop and tree species suggested by Kaburagi's experiments at Hitachi. These bridging objects, while valuable, were piecemeal in their creation and transmission. The absence of a coherent collection of bridging objects, however, did not retard policymaking. Government and company policies were generally successful in dealing with the worst aspects of the problem. This suggests that while a scientific problem-framework is essential for understanding subtle environmental problems, it is not necessary for blatant problems such as the smelter smoke problem of Meiji Japan.

In conclusion, the influence of scientists in the copper mining issue was oblique. Their legacy was less in influencing policy than in establishing an indigenous scientific capability to analyze environmental problems and setting precedent to feed this analysis into the policy process. This is the beginnings of Japan's "historical scientific momentum" on the acid deposition problem.

Sustainability

A complex set of interactions between domestic and international forces propelled Japan's unsustainable course in relation to acid deposition in the latter half of the nineteenth century. Readers may be surprised, as I was at first, how clearly and powerfully international forces shaped Japan's first experience with acid deposition problems. Imported scientific and technological knowledge were both sinner and savior. Japan's first acid deposition-related experiences—with smelter smoke and precipitation chemistry analysis—followed similar paths: international forces determined the initial trajectory of events but domestic forces took over and determined their final course.

Assuming that Japan was a sustainable society at the end of the Edo period, then whatever forces brought the Edo period to an end were also responsible for knocking Japan off of its sustainable track. The proximate cause for the end of the Edo period was the arrival of a foreign naval fleet. This event presented Japan with the distinct possibility of foreign domination or even colonial status. In other words, a perceived threat to national survival forced Japan out of its traditional, sustainable mode. To counter the threat to national survival, Japan embarked upon a massive program of industrialization and modernization. Japan imported Western science, technology, and institutions. One result was large-scale environmental (and acid deposition) problems, the worst of which were associated with the Ashio Copper Mine. Ashio in turn forced the sustainability issue upon Meiji society. Copper mine pollution set up a bitter social debate over (unsustainable) industry and (sustainable) agriculture. Industry was the savior in the face of foreign threat, but agriculture was the traditional wellspring of strength of the nation. Many saw peasant suffering due to pollution as sapping the spirit of the nation. In the end, the extreme distress caused by copper mines was deemed socially unsustainable, and strict remedies were instituted. It was not, however, deemed ecologically unsustainable in large part because the ecological cause-and-effect connections were unclear.

Thus, Ashio and the copper mining problem, while a social issue, never became an ecological issue.[28] Human health and economic well-being were the central issues. Nature was something to utilize for economic support and aesthetic pleasure, not something to be protected in its own right. The government finally acknowledged, in quasi-Confucian ethical fashion, a moral responsibility for the health and well-being of its citizens, and agreed there were limits to sacrifices that could be asked on the road to modernization. It also knew from its study of the Western experience that social disintegration would result if it remained indifferent. Eventually, a socially sustainable mode was reached in Japan by about 1920, as illustrated by the general lack of public protest after this date. Science was only indirectly responsible for defining this trajectory toward social sustainability, though. Why? Because science, in the terminology of this book, could not provide an authoritative solution path. Part of the problem was that it never got theoretically untracked to engage in the comprehensive cause-and-effect analysis needed to devise a solution path.

5

Period 2 (1920–1945): Flowering of Precipitation Chemistry

Between 1868 and 1945—the first environmental era of modern Japan—environmental problems were like tiny dust devils in a vast plain of industrialization. They were generally isolated episodes that flung up a lot of dust in a local area and then disappeared. None, with the exception of pollution from the Ashio Copper Mine, became a national issue. After Japan's early confrontation with copper mining-related pollution incidents between roughly 1890 and 1920, environmental problems never again reached center stage in Japanese politics until the 1960s. However, such problems persistently lurked in the shadow of industrial progress.

Early Urban Air Pollution

Although air and water pollution at copper mines was the most serious early form of environmental degradation, urban areas also suffered. Of all the cities in Japan, Osaka was the king of air pollution. The modern Japanese word for environmental pollution, *kōgai*, originated in Osaka (Hashimoto 1989, 7). It translates as "public injury or public damage" and was first used in an 1882 amendment to an 1877 ordinance ("Regulations on the Control of Manufacturing Plants") passed by the Osaka prefectural government. The purpose of the amendment was to include pollution control in factory regulation. The term reveals that very early in modern Japanese ecopolitical history pollution was singularly linked to human health and livelihood, not ecological impacts.

By 1911 Osaka was a forest of smokestacks. In that year it established a Soot and Smoke Abatement Committee (*Baien Boshikai*) headed by the prefectural governor; however, it disbanded in 1917 for lack of funds and interest (Iijima 1979, 47). In 1913 Osaka established the first urban air pollution monitoring program in Japan at the Osaka Public Health Laboratory (*Osaka Shiritsu Eisei Shikenjo*) (Hashimoto 1989, 7). As described in the last chapter, the first urban rainfall chemistry measurements in Japan seem to have been made at this institute in 1912. By 1920 the lab was measuring pollution via deposition into containers. Once a month the amount of undissolved substances and carbon (soot) that fell into large 55-liter glass bottles was analyzed. Acidity was not measured. The results were published in the laboratory's bulletin, *Osaka Shiritsu Eisei Shikenjo Jigyō Seiseki Gaiyō*.[1] However, despite the efforts to measure and control pollution, an Osaka newspaper declared in 1919: "Osaka, with its 1,947 stacks, must be the world's most polluted city" (Iijima 1979, 61).

Except for a brief interruption during World War II, air pollution research in Osaka has continued unbroken to the present. Sampling at thirteen sites in 1926 demonstrated that the volume of soot and dust in Osaka air was comparable to that of London, and by 1928 it was estimated that over 20,000 tonnes of soot were falling on the city each year (Iijima 1979, 70, 75). In the same year Osaka launched the nation's first Clean Air Week. Soon after, the term *hakuto* (smokeless city) was adopted as the city's clean air slogan, and in 1932 Osaka enacted Japan's first soot and smoke regulations, with modest success (Iijima 1979, 75, 81–90). Osaka's air pollution research and legislation strongly influenced other cities throughout Japan.

In 1927 Arimoto Kunitarō started an air pollution monitoring program in Tokyo similar to that in Osaka at the Tokyo Public Health Laboratory (*Tokyo-shi Eisei Shikenjo*) (Hashimoto 1989, 7–8; Iijima 1979, 75). Results were published in the laboratory bulletin, *Tokyo-shi Eisei Shiken Hōkoku*. In 1928 Tokyo sponsored a meeting of prefectural public health officers to discuss air pollution control and other sanitation issues. This seems to have been the first nationwide meeting at which urban air pollution was discussed. Although urban air pollution research began early, efforts at air pollution control were largely ineffective. During the 1920s Japan's political and industrial elite were busy with tasks other than worsening air quality. Japan became increasingly militaristic starting in the late 1920s, and by the 1930s the military had essentially taken over control of the national government. Environmental problems took a back seat to imperial expansion and war preparation. Despite this, Japanese scientists began in the mid-1920s an amazing burst of activity related to precipitation chemistry that lasted for about twenty years.

Precipitation Chemistry Comes of Age

As already described, the first measurements of the chemical composition of rainfall in Japan were performed by agricultural scientists. These measurements were few and geographically far between. It was not until the mid- to late-1920s that precipitation chemistry analysis, especially in urban areas, suddenly proliferated. Reconstructing the activity of this period is a little like trying to understand the behavior of a flock of wader birds after they have visited a tidal flat. There are footprints everywhere, but the movement of the whole is difficult to discern. The early history of precipitation chemistry is briefly recorded in several sources (see, for instance, Kurashige [1934], Miyake [1957], Fujita [1993a], and Maruyama et al. [1993]). However, these documents describe only the footprints, not the whole. Most of the history presented in this chapter was pieced together by sleuthing old journals and manuscripts.

Precipitation chemistry measurements by Akimoto Minoru of the Osaka Public Health Laboratory seem to be the first that directly address *acidic* pollution (Akimoto 1929). Akimoto describes precipitation analyses conducted from 1926–1928 at eleven sites around Osaka, which he refers to as "smoke city." His report contains the earliest statement I found of the link between acidified precipitation and industrial activity. It also includes an analysis of the seasonal and spatial differences in the data. Akimoto analyzed the pH of rainwater.[2] He found pHs ranging from 5.5–7.0. He also measured sulfate, chloride, ammonia, nitric acid, nitrogen dioxide, and other miscellaneous concentrations. The clearest correlation in his data is between sulfate (and acidity) and industrial areas. Rainwater was "pure" (pH near 7.0) in residential areas and acidic in industrial areas. Akimoto mentions that this finding is consistent with that of an English researcher who found rural areas to have slightly alkaline rain and industrial areas to have slightly acidic rain. One of the most curious measurements made by Akimoto was of the bacterial count in rainfall. He found a higher bacterial count in summer than winter. Of the eleven references in his article, all except for the one to the English researcher and one to the above-mentioned 1912 rainfall chemistry analysis in Osaka, are to German scientific literature.

It is not clear whether the type of research initiated by Akimoto was continued in Osaka. I found no reference to any further work in any major journals of the time. Other than Akimoto's work I found only one other reference—contained in Kurashige (1934)—to early urban rainfall chemistry analysis. Kurashige mentions in passing a 1931 publication by Mikuni Ryumon on rainfall chemistry in the city of Niigata on the Sea of Japan coast. This article was published in a local medical bulletin. Public health scientists do not seem to have actively continued precipitation chemistry work; it was soon taken over by meteorologists who then ushered in the true flowering of precipitation chemistry during the 1930s.

Prewar Research

1933: Matsudaira and the Kobe Observatory

The first meteorologist to analyze the chemistry of precipitation seems to have been Okada Takematsu, director of the Tokyo Central Meteorological Observatory (forerunner of today's Meteorological Agency). He also became director of the Kobe Meteorological and Oceanographic Observatory when it switched from a prefectural to national observatory in 1920. His work on precipitation chemistry does not appear to have been published, at least in analyzed form, for it is not referenced by later researchers. However, it was enthusiastically taken up by Matsudaira Yasuo and his colleagues at the Kobe Meteorological and Oceanographic Observatory. The earliest publications by Matsudaira date from 1933. Two articles, published in the official journal of the Kobe Observatory, *Umi to Sora* (Sea and Sky), which began publication in 1921, describe the acidity of snowfall and rainfall in Kobe (Matsudaira 1933; Matsudaira and Katō 1933).

The first brief article describes a "spontaneous" decision by Matsudaira and his coworkers to measure the chemical composition of snow that had collected on a large board just outside the observatory. A pH of 5.3–5.4 was obtained. Matsudaira compares this value to the normal pH values of rainfall obtained for Kobe of 5.9–6.4, as apparently measured by Okada. Thus, he concludes that snowfall is more acidic than rainfall. He then goes on to compare these results with those obtained by Cohen and Ruston around the industrial city of Leeds in England (Matsudaira references their book, *Smoke*). Cohen and Ruston found that in December in Leeds the precipitation was acidified due to sulfate ions. Thus, Matsudaira conjectures that it is industry around the Osaka-Kobe area that in winter is causing high sulfate values, and thus the low pH in the snow. It seems likely that the inspiration to measure snow pH came from acquaintance with Cohen and Ruston's work, although this is not explicitly stated in the article. Matsudaira concludes with the observation that deterioration of the white paint on the wall of the observatory facing the direction of Osaka was probably caused by sulfate ions blown from Osaka.

In a sequel to the first article, Matsudaira and his colleague, Katō Takeo, investigate the chemical composition of summer rainfall instead of winter snowfall. They state in the article that they decided to do this in their spare time. They rigged up a crude do-it-yourself collection system at the observatory. One of the authors also brought rainwater collected in a washtub at his home some four kilometers from the observatory. They then gathered and analyzed data from the end of July through September 1933, and found an average pH value of 5.6, with a range of 4.9 to 6.7. In addition to pH, Matsudaira and Katō measured other chemical species and collected typical meteorological data such as wind direction, wind speed, and rainfall amount. They found the pH was lower, and sulfate higher, when winds blew from Osaka toward Kobe. Again, the results were compared to those in England. Sulfate values in the 1930s in England were far higher than those in Kobe, on the order of two to ten times higher, according to a table in Matsudaira and Katō's paper. In addition, they compared their Osaka industrial activity/wind direction data with a similar analysis made in Leeds in 1906. The Leeds data showed that at a monitoring site outside of the city both sulfate and acidity significantly increased when winds blew from Leeds but dropped when they blew from a nonindustrial direction. Matsudaira and Katō state this is exactly the correlation they observed.

Matsudaira, like many of the scientists of this period, was not only prolific in the number of reports and articles published, but also prolific in the number of areas of research tackled. His research interests covered not only meteorology but also oceanography and limnology. Further articles related to precipitation chemistry include one on typhoon precipitation chemistry, specifically a typhoon that hit Kobe that had a pH of 5.8 (Matsudaira 1937); and one on the yellow sands blown from China, one of the earlier studies on the yellow sands phenomenon (Matsudaira 1938). As of 1995, long after his retirement, Matsudaira was alive and well and in his nineties.

1933: Yoshimura and Discovery of a pH 1.4 Volcanic Lake

Before continuing with the evolution of precipitation chemistry, we will pause for a brief introduction to the history of lakewater chemistry for lakewater chemistry and limnological studies developed parallel to and significantly influenced precipitation chemistry work.[3] We start with the discovery of what was at the time the world's most acidic lake. In 1932 Yoshimura Shinkichi discovered a small volcanic lake in northern Japan with a pH of 1.4.[4] The lake, Katanuma, is located at Naruko Hot Spring in an old crater on Mt. Katanuma in Miyagi Prefecture. The report of the discovery was published in English in the German limnological journal *Archiv für Hydrobiologie* (Yoshimura 1933). Katanuma also had a very high sulfate concentration of 474 mg/l. It was the most acidic of a series of acidic lakes that had been analyzed in Japan. The lakes were acidified by waters of volcanic origin rich in sulfuric and/or hydrochloric acids. Other lakes were reported (Yoshimura 1932) with pHs of 3.2 (Osoresan-ko in Aomori Prefecture), 3.7 (Onuma-ike in Nagano Prefecture), and 4.1 (Aka-ike in Fukushima Prefecture). However, what surprised Yoshimura most upon discovering Lake Katanuma's high acidity was the fact that it contained life. "Such a high acidity . . . as pH = 1.4 can not deny all the aquatic organisms!," he exclaims in his paper (Yoshimura 1933, 201).

Yoshimura was one of the most famous early limnologists in Japan. He worked at the Geographical Institute of Tokyo University. His first publications date from the 1920s, which eventually led to receiving the distinguished Japan Science Council award for outstanding contributions to science in 1935. In 1937 he published the first definitive textbook on limnology in Japan (Yoshimura 1937). In all his writings he shows a keen awareness and mastery of the latest research in Europe and the United States. In a review article on limnological research worldwide (Yoshimura 1942), he states that in the area of chemical analysis of lakewater Japan is one of the world leaders. Japan's research strength, he notes, was its comprehensive field research, not its contributions to theory. Yoshimura died at the age of only forty when in 1947 he fell through the ice of a lake where he was doing research.

As interesting as Yoshimura's discovery of the world's most acidic lake is, it is merely the tip of the iceberg of work done in Japan on lake pHs, lake chemistry, lake biology, and limnology in general. Yoshimura was not the first limnologist in Japan. That honor belongs to Tanaka Akamaro. Tanaka studied geography in Europe in the late 1800s while his father was ambassador to Italy. He returned to Japan in 1895 with the intent of studying lakes. Since there was no equipment for doing limnological research in Japan, he started from scratch and fashioned his own. Not only did he not have equipment but when he first tried to study lakes around Japan he was often prevented from doing so by superstitious locals who were afraid that his doings in the middle of a lake would anger the lake spirits. He persevered, though, and became the foremost limnologist and teacher to Japan's first generation of limnologists. One who followed in Tanaka's footsteps was Sugawara Ken, the first chemist to study lakes. We will have occasion to meet Sugawara at the end of this chapter. One of Sugawara's students was Yoshimura.

Neither Yoshimura's research nor that of his contemporaries, to my knowledge, ever made a link between acidic inputs from the atmosphere and changes in lake chemistry, although they did make the more obvious link between acidic waters of volcanic origin and lakewater acidification. It is conceivable that an atmosphere-lakewater link related to acids could have been made, especially in relation to volcanic activity, because there was frequent communication between meteorologists, oceanographers, volcanologists, and limnologists in Japan if one is to judge by the cross-referencing in journal articles. Matsudaira's work, for instance, is referenced by Yoshimura, and visa versa. But the link seems to have gone unexplored.

1934: Kurashige and the Early State of Precipitation Chemistry

In 1934 a short review article on Japanese precipitation chemistry up to that time appeared in the first volume of a new journal titled *Tenki to Kikō* (Weather and Climate) published by the Central Meteorological Observatory in Tokyo (Kurashige 1934). The author of the review, Kurashige Eijirō, describes then current precipitation chemistry research. Unfortunately, much of the information is vague and has the flavor of insider's shoptalk. In describing pH results around the country, Kurashige includes his own preliminary results for Tokyo, which he obtained the previous year (1933), and which were published in greater detail one year after the review (Kurashige and Kagei 1935). Kurashige's data seems to have been the first on precipitation chemistry in Tokyo, for he references no earlier works. The pH values of precipitation chemistry for various cities around Japan are listed in table 5.1.

Kurashige draws several interesting conclusions about the acidity of rainfall in Japan in his two papers. First, he notes that in Kobe when winds blow from Osaka very low pH values were observed. This is, as we have seen, from Matsudaira and Katō's work. Second, Kurashige concludes that Osaka is the most polluted city in Japan, and that Tokyo is second because Akimoto's sulfate measurements in Osaka are higher than his for Tokyo. Third, based on his work in Tokyo he deduces that the city center is more polluted than the suburbs. An average pH of 5.4 was found for the city center, whereas the suburbs had a value of 5.8. Fourth, again based on his own work, he states that precipitation in

Table 5.1
pH values in cities in Japan in the early 1930s

Location	pH Range
Osaka	5.4–7.0
Niigata	5.6–6.9
Fukuoka	5.4–7.2
Kobe	4.9–6.7
Tokyo	4.6–6.2
Northern Japan (snow)	5.8

Source: Kurashige (1934, 391), and Kurashige and Kagei (1935, 215).

Tokyo was strikingly acidic at the beginning of a rainfall and became progressively alkaline over the course of the rainfall. Fifth, in Tokyo high acidity was observed during the daytime when "big city activity is at its greatest" (to paraphrase Kurashige's Japanese expression), but not during the nighttime. In his 1935 paper he notes that by this time not only were precipitation chemistry measurements being made in Osaka, Kobe, Fukuoka, and several locations in Tokyo, but also that snow chemistry analyses were being conducted in twenty locations in Hokkaido, Aomori, Yamagata, and Niigata prefectures. Clearly precipitation chemistry research was taking off.

1935: Systematic Chemical Analysis of Precipitation

It seems that the previously mentioned Okada Takematsu was the first meteorologist to see the importance and necessity of conducting systematic chemical analyses along with traditional meteorological measurements (see discussion in Maruyama et al. [1993]). Due to his urging, a Chemical Analysis Section of the Central Meteorological Observatory in Tokyo was established in 1935. Miyake Yasuo was appointed its first director.[5]

One of Miyake's first tasks was to review methods of precipitation chemistry analysis. This he accomplished over a five-year period beginning in 1936 in a multipart series titled "A Critical Study on the Methods of Water Analysis," published in *Kishō Shūshi* (Journal of the Meteorological Society of Japan), the official journal of the Tokyo Central Meteorological Observatory. In these articles, Miyake reviews, comments on, and discusses improvements for the then state-of-the-art methods of water chemistry analysis (including chloride, sulfate, phosphate, silicate, nitrite, and nitrate analysis). The articles all show the strong imprint of European, particularly German, and American chemical studies.

Under his leadership the Chemical Analysis Section launched in 1935 a program of systematic chemical analysis of precipitation at national meteorological observatories in Tokyo, Hamamatsu, Utsunomiya, Miyako, Chichijima, and Kobe. (The results obtained at some of these stations will be discussed later.) Precipitation chemistry analysis continued at these and other stations over the next twenty-five years until 1961 when they were discontinued. Ironically, except for some of the initial data discussed later in this chapter, the data sets collected by the Chemical Analysis Section were never analyzed. It just sat as raw data. The first analysis of a complete data set—the 1935–1961 data set for Kobe—was published in 1997! This is discussed in the next chapter.

According to the Central Meteorological Observatory's 1935 *Kishō Yōran* (Meteorological Survey), where raw data was published, precipitation was collected in a series of open metal pots set on stands on observatory rooftops. The pots were washed with distilled water before a forecasted rainfall event to minimize contamination by dust and soot. A rain gauge was set up near the collector. The following items were measured: rainfall amount, pH, Cl, SO_4-S, NO_3-N, NH_4-N, and NO_2-N. Data were reported for each rainfall event.[6]

1935: First Dry Deposition Measurement

Katō Takeo, the Kobe Meteorological and Oceanographic Observatory researcher we met earlier, published in 1935 a highly innovative piece of research on measurement of acidic dry deposition (Katō 1935). This is the first dry deposition measurement in Japan. Later, he repeated and improved his measurements and published them in another article four years later (Katō 1939). Katō's method was simple. He washed off dust and particles attached to pine needles taken from selected locations in the port city of Kobe (twelve in the first study, and twenty-five in the second) with distilled water, and then measured the chemical composition of the water. He himself states that this was the first time that such an analysis had been carried out, and laments the fact he was not able to do it as perfectly as he would have liked (he did it in his spare time). Despite the imperfect method, he produced the first isopleth maps of pH and sulfate ion distribution in Japan (see figure 5.1). The maps are, to my knowledge, the earliest isopleth maps of any kind related to acid deposition in Japan. He found a range of (dry deposition) pHs of 4.3–5.2 in the first study, and 4.4–6.4 in the second. He found that low pH isopleths correspond to the port areas where the industrial belt of Kobe was located. He notes that these values were somewhat lower than the values for rainfall found by Matsudaira and himself earlier. Katō does not refer to his work as a dry deposition analysis, he called it "analysis of soot attached to leaves."

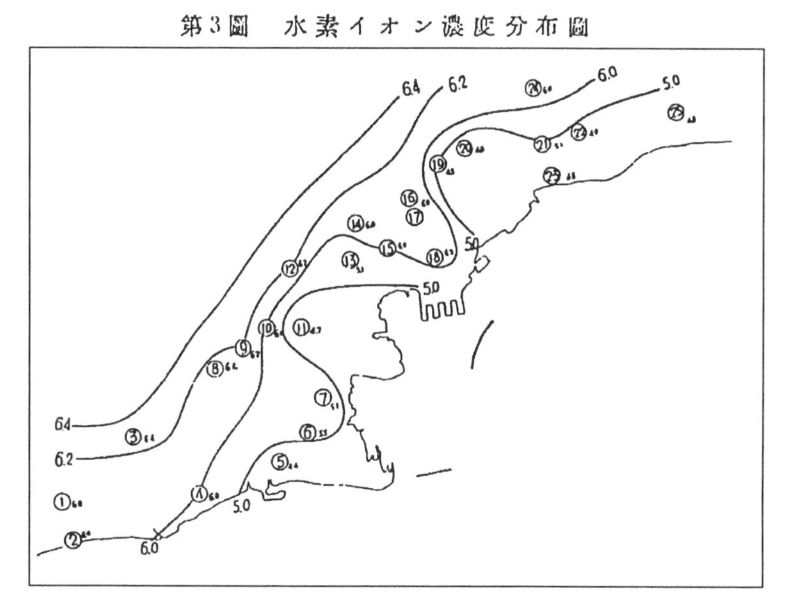

Figure 5.1
pH isopleth map for Kobe in the 1930s. Reprinted from Takeo Katō, "Ko no Ha ni Tsuku Ojin no Kagaku Seibun Kekka nitsuite—Dai ni hō" (Chemical Analysis of Soot Attached to Leaves—Part 2), *Umi to Sora* (Sea and Sky) 19, no. 10 (1939): 305, figure 3. Used with permission of *Umi to Sora*, Marine Meteorological Society, Kobe, Japan.

1939: Miyake and Urban Precipitation Chemistry

The first comparative analysis of urban precipitation chemistry in Japan was published in a now-famous paper titled "Usui no Kagaku" (The Chemistry of Rain Water) (Miyake 1939). Miyake synthesized, summarized, and compared precipitation chemistry data for several urban areas. By this time precipitation chemistry analyses were being carried out in many urban locations. Raw data were being published in the Central Meteorological Observatory's *Kishō Yōran*, and analyses were being published in journals such as *Kishō Shūshi*, *Umi to Sora*, and *Tenki to Kikō*. Miyake performed a comparative analysis of the 1936–1938 data sets for three cities: Tokyo, Kobe, and Hamamatsu. Preliminary results for Tokyo had already been published by Kurashige in 1934 and 1935, for Kobe by Matsudaira and Katō in 1933 and 1935, and for Hamamatsu by Shimizu (1936). Miyake's analysis, however, used only "standardized" data obtained after the Chemical Analysis Section established specific collection and analytic techniques. Miyake determined monthly and yearly averages for such species as Cl^-, SO_4^{2-}, NO_3^-, NH_4^+, Na^+, K^+, Ca^{2+}, Mg^{2+}, and H^+ (pH), and studied the relationship between several of these species. Only a portion of his results are discussed here.

Miyake found a number of interesting trends related to acidity. He showed that the monthly average of sulfate was highest in winter (November through April) and lowest in summer (March through October). He suggested that this was due to the use of coal and charcoal for winter heating (at this time coal and charcoal were the most common residential heating fuels in cities). This seems to be the first time the link between winter domestic coal use and high sulfate concentrations was postulated in Japan. Miyake also found that the yearly variation of sulfate in rainwater correlated with that of sulfate in the air. Thus, he postulated that sulfate in the air that was being incorporated into the rainwater. Monthly and annual average pH values were also reported. He gives the three-year average (1936–1938) pH for the three cities as: Tokyo (4.1), Kobe (5.2), and Hamamatsu (5.6). His data also show that for Tokyo the pH values were lowest in winter and highest in summer, although the monthly variation was not as striking as that for sulfate, suggesting to him that other factors were influencing the pH.

Miyake's paper presents a clean technical synthesis of raw observational data, and was the first in Japan to compare urban rainwater chemistry using standardized techniques. Although, as already described, others preceded him, Miyake established the first firm synthesis of scientific knowledge on urban precipitation chemistry. His work was referenced time and again in the next decades, and was the foundation upon which others elaborated. In addition, Miyake's article would, much later, provide historical perspective when the problem of acid rain was rediscovered in Japan some thirty-five years later. Miyake did not raise a prophetic clarion call warning the nation of impending acidification of the urban atmosphere, though. Nowhere did I find any statement that he construed his work as uncovering a significant problem. He did not seem to view the low pH or high sulfate values in Tokyo, for instance, with concern.

Prior to his work on rainwater chemistry, Miyake published one of the first chemical analyses of urban ambient air pollutant concentrations in Japan titled "Atmospheric Impurities in Central Tokyo" (Miyake 1937). He begins the article with the statement that though elsewhere (Europe and America) awareness of, and research on, the chemistry of trace substances in the atmosphere was growing, in Japan the dawn of such research had not yet broken (to paraphrase his figure of speech). To remedy this situation, in March 1937 he began in Tokyo an investigation of trace substances. In his analyses of sulfur (SO_2 and SO_4) concentrations in Tokyo, Miyake compared his results to a report on air pollution published in London the same year. He notes that the sulfur concentrations in London air were vastly higher than those of Tokyo. It is reasonable to surmise that this is one reason he shows no particular alarm about Tokyo's atmospheric acidity—he was comparing Tokyo air to the air of what was almost certainly the dirtiest city in the world at the time, London. The reference to the London air pollution report (which he only mentions in the text, but does not formally reference) clearly shows that Miyake was acquainted with foreign research on air pollution.

In a second article (Miyake 1938), he reports the results of a one-year compilation of air pollution data (the first article contained the results of only four months of data). For sulfur concentrations, the results showed that in autumn the "impurities" were lowest and in winter highest. Again, similar to the first article but now with more data to back it up, he concludes that the air in Tokyo is much cleaner than in London. After publication of this second article, Miyake does not seem to have personally continued research on ambient air pollutant concentrations. Instead he shifted to precipitation chemistry, the first result of which was the previously described 1939 article. However, his work on air pollutant concentrations was continued by Matsui Hideo, who published two further articles on "atmospheric impurities" in Tokyo (Matsui 1939, 1941).

After the war, Miyake's research falls out of the mainstream of our history of acid deposition science. But before we leave this dynamic individual, a few remarks will be made on the rest of his career. After the war his work turned away from urban air and rain pollution to the geochemistry of the ocean and atmosphere.[7] His postwar papers cover a vast range of subjects—ozone formation in the atmosphere, ultraviolet radiation during a total eclipse, radioactive fallout in Hokkaidō, the chemistry of rivers in Japan, chemical analyses of volcanic ash, carbon dioxide contents in the atmosphere and oceans—to mention a few. He wrote two books—*Earth Chemistry* (Miyake 1954), which was translated into English in 1965 and into Russian in 1969, and *Precipitation Chemistry* (Miyake 1957). Miyake's main scientific interest after the war was "water and Earth chemistry," and, as he states in a paper in 1969, through a lifetime of water-related studies he came to realize water was a "little universe" and that the more he studied this little universe the less he knew about it (Miyake 1969, 16).

However, rather than his water studies, Miyake in the 1950s became internationally famous for his scientific and personal interest in environmental radiation and the nuclear fallout problem.[8] The galvanizing event that motivated him to devote a major fraction of his abundant energy to radiation studies was the tragic story of the *Lucky Dragon* and the first hydrogen bomb test by the United States on Bikini Atoll in 1954.[9] As a result of this tragedy, Miyake designed and personally led research efforts throughout Japan and the Pacific to study the effects of radiation on the environment. He founded the Japan Radiation Research Society in 1959, and the Peace Society of the Lucky Dragon about the same time. He repeatedly stressed that scientists have a responsibility to prevent nuclear war. He was active in this and other diverse research right up until the time of his death in 1990.

1940: Japan's First Acidified Lake

Returning to our acid deposition history, several decades before the acidification of Scandinavian lakes drew worldwide attention to the acid rain problem, Japan itself experienced the effects of unintentional acidification of a lake (Murano 1993, 119–121). As part of Japan's war effort, attempts were made to increase hydroelectric power production. In one such attempt, one of Japan's most unique and fabled lakes was accidentally acidified. Lake Tazawa in Akita Prefecture in northern Japan is the deepest lake in Japan (depth 424 meters) and was long famed for a unique species of trout known as *kunimasu*. However, to increase power generation from a river exiting the lake, officials decided to raise the lake level.

Therefore, in 1940 a nearby acidic river with a pH of around 3.0, which originated in a volcanic hot spring of pH 1.1, was diverted into the pH neutral Lake Tazawa. Within a few years the lake became acidic; it dropped to a pH of 4.7. Acid sensitive life-forms began to die off, and before long the *kunimasu* trout at the top of the food chain went extinct. This seems to be the first incident in Japan of human-induced acidification of a lake. In 1989 a "neutralization plant" was built to add lime to the lakewater at a cost of U.S.$32 million (with an annual operating cost of about U.S.$2 million). As of 1993 the pH of the upper portion of the lake had come up to 5.1, but the lower levels of the lake were as yet unchanged.

Research during and after the War

One year after Miyake's famous 1939 article another piece of research was published by a colleague on the precipitation chemistry of Utsunomiya, a town in the Kantō Plain about 100 km north of Tokyo (Ogura 1940). The author, Ogura Yutaka, mimics Miyake's format and begins the article with the comment that the study of rainwater chemistry has suddenly taken off in Japan. Ogura proceeds to analyze a 1938–1939 data set from the meteorological station in Utsunomiya. He found an annual average pH value of 5.5. He notes that the sulfate, nitrite (NO_2^-), and nitrate (NO_3^-) concentrations increase in winter and decrease in summer, and concludes with the interesting observation that the nitrite concentration in rainwater during lightening storms was about twice that of ordinary rain. Ogura does not reference any foreign works (Miyake referenced many); however, he acknowledges his indebtedness to Miyake.

One year later yet another article was published by Hirasawa Kenzō analyzing rainwater in Miyako, a small fishing town in northern Japan (Hirasawa 1941). The town, backed by mountains and hugging the rough Pacific coast approximately 500 km north of Tokyo, recorded an annual average pH value of 5.5. Again, the format and style of

Hirasawa's analysis is almost identical to Miyake's original paper. The author states he is following the same methods as Miyake, and also acknowledges the support and previously mentioned work of Ogura. Although the author mentions that a smelter was recently built (the only large industrial plant in the town), it does not seem to have acidified the atmosphere because of its enormous Hitachi-like stack.

Matsui Hideo, who took over Miyake's work on air pollution in Tokyo, published another in the series of "atmospheric impurities" articles; this one titled "Atmospheric Impurities in the City of Maebasi" (Matsui 1942a). It analyzes air pollution in Maebashi, a town in the Kantō Plain about 100 km northwest of Tokyo. The analysis was for only a ten-day period in July 1941 during which time, among other results, he found an average rainfall pH value of 4.4. Also in 1942, Matsui published a sequel article to Miyake's original 1939 article, aptly titled "The Chemistry of Rain Water—Part II" (Matsui 1942b). The article analyses precipitation chemistry in Tokyo and Maebashi.

In the midst of the now-raging Pacific War, an interesting one-month piece of research was done in Kyūshū in 1943. The researcher, the previously mentioned Matsui, measured the chemical composition of river fog. This is quite likely one of the first investigations of the acidity of fog in Japan. There is no mention as to why this particular research was conducted or why this location was selected, but it was almost certainly connected with the war effort and was surely done in Matsui's spare time. Although the results do not appear to have been published at the time, it is mentioned in an article by Miyake after the war (Miyake 1950). A remarkably high sulfate concentration (225 mg/l) was found in the river mist. Matsui speculates it was from steam engines running along the river and from homes and factories located near its banks.

Not long after Matsui's river mist measurements, the war entered its final phases and the normal pursuit of science in Japan became all but impossible. On 10 March 1945 General Doolittle's firebombing of Tokyo took place. On 6 August the atomic bomb was dropped on Hiroshima. And on 2 September Japan unconditionally surrendered. As Japan entered the road to economic and social recovery after the war, Japanese scientists picked up as best they could where they had left off. Articles on precipitation chemistry began to reappear.[10]

1948: Sugawara and the Pure Chemist

Almost immediately after the war, a highly sophisticated piece of acid deposition research appeared. The author was Sugawara Ken of Nagoya University, who we met earlier as the first scientist to study lakewater chemistry in Japan. His article (Sugawara 1948), titled "Precipitation Chemistry," was published in the Japanese journal *Kagaku* (Science). The research seems to be the first in which he studied precipitation for he references no earlier works under his authorship. Sugawara begins the article with the statement he is a "pure" chemist tackling the problem of rainfall composition and formation, and that few pure chemists tackle such "practical" problems. He continues with the acknowledgment that precipitation chemistry in Japan is already some twenty years old, and that its founding fathers are Miyake and Matsudaira. He goes on to emphasize that the major practical application of the type of research he was doing is the creation of artificial rain and snow. This work, he notes, was already being done in America. Most of the references in the article are to American and German literature.

The bulk of the article describes field observations over the two-year period from 1946–1947 in the Nagoya area on the chemistry of rainwater, with particular attention to the influence of the sea. Included in the article is what is probably the second analysis of dry deposition in Japan. (The first was that of Katō of Kobe, which Sugawara does not mention. Sugawara's analysis was more refined, and more clearly differentiates dry and wet deposition.) Sugawara compared the chemical composition of rainfall and of substances dry-deposited on pine trees in three locations around the Nagoya area: (1) on an island in the sea not far from Nagoya, (2) in the city of Nagoya itself, and (3) in a remote inland town in the Japanese Alps north of Nagoya.

To analyze dry deposition, needles on pine trees were washed with distilled water. Then, after several days of clear weather, they were again washed with distilled water, and the wash water collected and analyzed. In the paper he does not compare the dry and wet deposition values, although it is easy to see from his data that dry deposition of sulfate was: (1) one-half that of wet deposition on the island, (2) double that of wet deposition in Nagoya, and (3) roughly equivalent to wet deposition in the mountain area. It is worth noting that I have run across no reference to these dry deposition results, or to Katō's analysis of "soot clinging to leaves," in later papers right up to the present, although Sugawara's wet deposition analysis is often mentioned in later works. It seems that this work on dry deposition was considered subsidiary to wet deposition analyses, not a field of study in its own right. The measurement that is most often referred to in later literature is the finding that the annual average SO_4^{2-} concentration in precipitation for the city of Nagoya in 1946–1947 was 0.34 mg/l. This is an extremely low value compared to those later measured in large cities in Japan, and reflects the low level of industrial and commercial activities in the immediate postwar period.

1950: End of the Flowering

(Note: Even though some of the material in this chapter spills somewhat beyond 1945, I chose the chronologically convenient date of 1945 to end the second acid deposition period; the transition between the second and third acid deposition periods is not sharp). Miyake (1950) published a short review article on precipitation chemistry in which he stated that Japan was one of the most advanced countries in the world in the field of precipitation chemistry. Unfortunately, even though he lauded the significant Japanese achievements in precipitation chemistry between 1930 and 1950, he does not offer explicit evidence for the claim (one has to take his word as the foremost Japanese expert in the area). Japan by this time possessed a considerable wealth of precipitation chemistry data and over twenty years experience in collecting and analyzing such data. Despite this outstanding start, after around 1950 the pursuit of precipitation chemistry knowledge related to acidic substances in the atmosphere passed into a quiescent phase, out of which it did not emerge until the mid-1970s. For the next twenty-five years, ambient air quality issues dominated. Ironically, Miyake's optimistic and laudatory review article can be seen as symbolically marking the end of the first great flowering of precipitation chemistry research in Japan.

Analysis of Period 2

Beginning around 1920 the foundation for Japan's present-day acid deposition science was established, and with it the humble beginnings of an acid deposition problem-framework and expert community. Acid deposition science, as a distinct field of inquiry, initially established itself in Japan in the area of precipitation chemistry that, as we have seen, blossomed in the 1930s and early 1940s. Researchers such as Matsudaira and Katō of the Kobe Meteorological and Oceanographic Observatory, Miyake, Kurashige, and Matsui of the Tokyo Meteorological Observatory, and Sugawara of Nagoya University were among the early pioneers, with Matsudaira and Miyake the acknowledged leaders. At the same time that precipitation chemistry was blossoming, the related field of freshwater chemistry also flourished. The most famous early figures were Sugawara and his student Yoshimura. Japanese scientists compiled a wealth of data on precipitation and lakewater pH and chemical composition by 1945.

An epistemic shift from a crude "copper smelter smoke" problem-framework to a more sophisticated but still ill-defined problem-framework centered around "urban precipitation chemistry" distinguishes Japan's first acid deposition period from its second. The lessening of smelter smoke problems and the worsening of urban air quality prompted the shift, and importation of Western scientific ideas related to urban air pollution, including urban precipitation chemistry, provided its intellectual momentum. Despite the rapid growth in knowledge during the second period, acid deposition science never intruded into politics. Even though scientists identified acidified urban precipitation (e.g., pHs of around 4.0 in Tokyo), they never singled it out as a distinct problem with political implications. The non-interaction between acid deposition science and politics is illustrated in the truncated flow diagram in figure 5.2. Key questions related to this period are: (1) how and why was Japan's acid deposition (precipitation chemistry) research program established, (2) why didn't Japanese scientists "discover" an acid deposition problem and construct a solid problem-framework, and (3) why didn't scientific knowledge serve as a launchpad for political action?

Establishing Acid Deposition Science

During Japan's second acid deposition period, precipitation chemistry was pursued and systematized in a way that never occurred during the first period. Why? A combination of reasons explain the change.

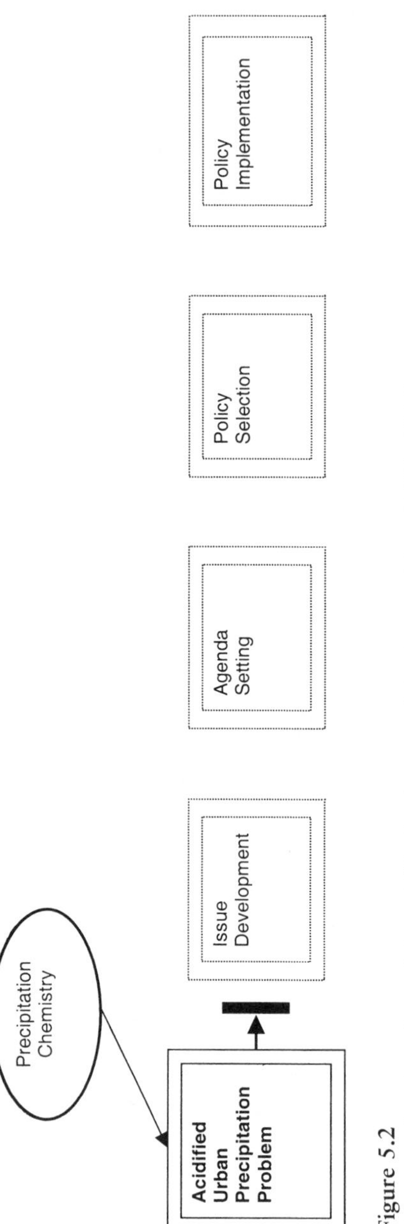

Figure 5.2
Science-policy interface of the acidified urban precipitation problem (1920–1945)

First, as already discussed in the last chapter, the basic character of science itself changed between the first and second acid deposition periods. The role of the scientific researcher as we know it today was more or less established during the second period. As compared to the first period, scientists during the second period were freer to engage in purer forms of research, such as precipitation chemistry, even though the research seemingly had no immediate, practical application. Hence, rather than research being directed toward solving a specific problem (such as smelter smoke), much of the research during the second period (such as on precipitation and lakewater chemistry) contributed to an understanding of natural phenomena for its own sake. The shift of the locus of precipitation chemistry data-gathering from agricultural scientists during the first period, to public health specialists during the first and early second period, and finally to geoscientists during the bulk of the second period illustrates the new emphasis on pure research.

Practical applications of precipitation chemistry research were not forgotten altogether, though. A second reason for the rise of precipitation chemistry research was its connection with urban air pollution. Air pollution was becoming a serious problem in urban areas and analysis of precipitation chemistry was a means of indirectly analyzing air pollutants. There was keen awareness of foreign air pollution-related work, especially that in Europe. Matsudaira refers to Cohen and Ruston's work on air pollution around Leeds, England. Miyake refers to reports on London's air pollution. It can be inferred from the Japanese literature that awareness of this type of foreign work inspired and prompted similar analyses in Japan. It is noteworthy, though, that I found no statements in any of the articles I surveyed attributing indigenous research directly to foreign inspiration. There were, however, many acknowledgements of fellow Japanese scientists' work. One can speculate that foreign inspiration was so commonplace that it didn't need reiterating in text.

A third reason for the rise of precipitation chemistry research, one actually mentioned by many researchers in their papers, was to study the effect of the surrounding seas on Japan's atmospheric chemical composition. Given the magnitude of the sea's influence on the Japanese landscape and culture, it is not surprising that the sea-atmosphere relationship was studied as a natural phenomenon in its own right.

Whatever the combination of motivating reasons, it is clear that the approach for studying precipitation chemistry was acquired from abroad, primarily Europe. Japan initially received its methodologies, its universal knowledge base, and its orientations to research from the West. As part of this influx, Japanese scientists acquired a vague but discernable problem-framework related to acidified urban atmospheres. Unlike the first acid deposition period, the means by which the methods, knowledge, and orientations were transferred to Japan were dominantly indirect (via scientific publications) not direct (via foreign teachers and overseas study). The Western (primarily European) framework was absorbed piecemeal, not in a single lump. As an example of how it was transferred, Miyake's first task as director of the Chemical Analysis Section was to codify for the first time in Japan precipitation chemistry analytic techniques. In order to do this, he drew almost exclusively from European sources. Miyake's achievement was a key scientific step to making pollution-related precipitation chemistry the object of full-time study in Japan.

After initial absorption of the European framework, a process of "indigenization" began. In other words, the universal techniques developed in the West were adjusted, modified, and applied to the Earth space/time-specific conditions of Japan in order to acquire a Japan-specific knowledge base (of pH values in cities, for instance). The initial localization process took place at national level, elite institutions (such as the Tokyo Meteorological Observatory) and from there diffused to local institutions. By the end of the second period, Japan possessed a solid, localized knowledge base. Japanese scientists had fully absorbed the Western techniques of the time and could express confidence in their accomplishments in the fields of precipitation chemistry and lake chemistry.

Both Miyake in precipitation chemistry and Yoshimura in limnology stated in review articles that Japan was among the world leaders in field observation (but not theoretical analysis). Similar assertions could probably also be made for many other fields of research that eventually became associated with the modern acid deposition problem, such as soil science and forestry, but I did not investigate these areas. All in all, Japan began to have confidence in its ability to conduct environmental science.

However, Japan's acid deposition problem-framework at this time is perhaps best described as "incipient." It was better defined than that of the first period, but it still lacked a clear theoretical structure. Precipitation chemistry data rapidly accumulated between 1920 and 1945 leading to the creation of indigenous methodologies and knowledge bases, but the Japanese framework was little expanded beyond its received boundaries. For example, no theoretical connection seems to have been made between the fields of precipitation chemistry and lakewater chemistry in relation to acidity even though acidity was measured in both fields.

An equally fuzzy Japanese expert community devoted to analysis of pollution-in-precipitation grew along with the problem-framework. My best guess, based on the number of researchers who authored journal articles or were mentioned in articles, is that it consisted of about a dozen or so individuals at the national and local levels, all of whom pursued this type of work part-time. Based on cross-referencing in journal articles, this group seemed to know each other quite well, providing further evidence of its "community" status. Tiny as it was, this band constituted Japan's first identifiable expert community related to acidification of the atmosphere. It exercised an influence all out of proportion to its numbers, though, for it laid the foundation not only for future definition of the acid deposition problem but also for future expansion into a true acid deposition expert community.

Three of the central figures mentioned in this chapter—Miyake, Yoshimura, and Sugawara—are often referred to by contemporary scientists as being among the founders of environmental science in Japan.[11] They were the first generation of scientists to consciously contribute to Japan's environment-related scientific knowledge base. It was their students (the second generation) and their student's students (the third and present senior generation), who continued the work. Although Japan's acid deposition-related research was largely patterned after Western research, the Japanese expert community succeeded in establishing a native capability and confidence whose importance cannot be underestimated. This capability was established long before any other country in Asia, and is one reason for Japan's leadership in acid deposition science in Asia today and for the "historical scientific momentum" I detected in my study of the present transboundary air pollution issue in East Asia.

Acid Deposition Issue Goes Unrecognized

Establishment of acid deposition science did not lead to recognition of acid deposition as a political issue. Why? What was missing? In a word, a clear-cut problem-framework from which to formulate well-grounded bridging objects was missing. Research on pH and sulfate concentrations in rainfall almost invariably mentioned their relationship to air pollution. However, this did not lead to singling out acidification of the (urban) atmosphere as a stand-alone problem despite the fact that many scientists conjectured about the relationship between industrial activity, coal burning, and high acidity.

Matsudaira inferred in 1933 that Kobe's low pH values occurred when winds blew from Osaka. Katō's isopleths of pH and sulfate for Kobe in 1935 show that the industrial belt of Kobe was the source of the lowest pHs and highest sulfate concentrations. Miyake speculated that the winter high sulfate/low pH values in Tokyo of the 1930s were due to residential burning of coal and charcoal. Despite these observations, no one spelled out a broad working hypothesis about the relation between industrial emissions, ambient air pollutants, and acidified precipitation, and then set about testing it.

Ironically, Japanese researchers had many rudimentary pieces for constructing a plausible problem-framework. They had a rough idea of cause and effect (e.g., the industrial activity, wind direction, peeling paint hypothesis of Matsudaira and Katō at Kobe); they had a handle on the extent of the problem (e.g., data from the network of urban stations set up under the auspices of the Chemical Analysis Section); and they had a sense of impacts (e.g., Kaburagi's older research on plant damage near smelters). However, the pieces of the puzzle were never put together.

We can speculate that Japanese scientists did not single out acidification of the urban atmosphere as a problem worthy of framework construction because, first of all, the imported scientific knowledge from the West did not emphasize it as a distinct problem. Thus, Japanese researchers were not on the lookout for it. This a good example of the "orienting power" of imported knowledge. Second, against a backdrop of increasingly severe ambient air quality problems, acid deposition did not stand out. Thus, Japanese researchers overlooked it. And third, the impact aspects of acid deposition science were not tightly linked to the meteorological. Knowledge of impacts gained during the smelter smoke period does not seem to have been deemed relevant to analysis of acidification of the urban atmosphere. This stunted problem-framework construction, and hence it remained incipient.

The lack of a well-defined problem-framework inhibited the formation of science-to-policy bridging objects. Japanese scientists admirably utilized the universal science-to-science bridging objects when adapting Western acid deposition-related knowledge to Japanese soil, but the lack of an indigenous problem-framework prevented development of bridging objects that might have influenced the policy process in Japan. There were no cause-and-effect, extent-and-intensity, or impacts bridging objects to trigger development of a political issue. Thus, there were no conceptual devices that political actors could employ to calculate their interests and determine their actions. As compared to the first acid deposition period, the type of acidification problem being experienced in the second period was subtler. It was not as blatant as smelter damage. Since scientists were unable to silhouette the problem for policymakers, no action was taken.

Be this as it may, precipitation monitoring in and of itself was almost as advanced in Japan as in Europe. It was perhaps more advanced than North America. Starting in 1935 Japan established a network of some twelve stations under the Central Meteorological Observatory that analyzed precipitation chemistry. This was earlier than even the first precipitation chemistry network in Europe.

Sustainability

Sustainability was not an issue during Japan's second acid deposition period even though the industrialization process in Japan intensified. Pollution problems, especially in urban areas, were bad, but at no time during this period did a crisis erupt such as the copper mines crisis in the early part of the century. Over and above the government's higher priorities (military and imperial ambitions dominated national decision making) and citizens' general willingness to put up with the downsides of industrialization, environmental problems switched between the first and second periods from being associated with single point sources in rural areas (copper mines) to being associated with multiple and diffuse sources located in urban areas. Because the sources were smaller and more numerous (e.g., residential coal-burning stoves), it was difficult to focus political attention on their individual impact. The shift from a rural to urban venue meant that the "agriculture versus industry" debate that dominated the sustainability question during the first period was missing. Also, the urban working and middle classes were still developing in Japan, and they had not yet found their political voice.

Using the terminology of this book, the incipient problem-framework of the second period lacked a "solution path" to give the problem policymaking direction. Such a path could have been formulated without a clear outline of physical and chemical cause-and-effect relationships and without knowledge of the exact extent and intensity of the problem, but it could not be formulated without a clear sense of the impacts on society. And this essential ingredient was missing.

The key development related to Japan's sustainability during the second period, as exemplified by precipitation chemistry research, was the indigenous building of scientific knowledge and capacity. As in the first period, Japanese scientists continued to import techniques and knowledge from the West; however, the means of import became indirect not direct, and greater effort was expended on internalization of what was imported. Japan's domestic scientific capacity building was truly impressive. For instance, the quality control exercised over precipitation chemistry monitoring data almost matches modern standards. Even though the acid deposition-related knowledge base lacked a theoretical foundation, the fact that Japan could conceivably have discovered, based on its precipitation chemistry network, large-scale acidification of the atmosphere long before it was discovered in Europe is high praise in and of itself. The flowering of precipitation chemistry that characterizes Japan's second acid deposition period actually represents a flowering of domestic environmental scientific capacity. And solid domestic capability in environmental science, as Japan and the world would soon find out, is a cornerstone of sustainable societies.

6

Period 3 (1945–1974): Air Pollution Reigns

Second Environmental Era (1945–1967): Postwar Reconstruction

After the devastation and defeat of World War II, Japan began a reconstruction process within the framework of a significantly altered political and social system. Industry was rapidly rebuilt, double-digit economic growth and promises of prosperity became the nation's watchwords, and the environmental mind-set (or lack thereof) that prevailed before the war continued unaltered after the war. By the mid-1950s, though, hideous industrial pollution began to rear its head again.[1] The scale of postwar industrialization soon far surpassed that of the prewar period, and with it came pollution problems on a scale unprecedented in Japanese history. In particular, air pollution in Japan's large cities became among the worst in the world. Acid deposition, though largely unnoticed, followed in its wake.

Japan's third acid deposition historical period (1945–1974) spans two environmental eras: postwar reconstruction (1945–1967) and the domestic environmental revolution (1967–1973).[2] Since air pollution was a central element of these two environmental eras, and since air pollution problems were the background for Japan's slowly evolving acid deposition science and policy, select aspects of the air pollution crisis between 1945 and 1973 are highlighted in the first part of this chapter. This sets the context for the quiet but continuing development of acid deposition science in the second part of the chapter.

Air Pollution's Climb: 1950s and 1960s

The first postwar response at the local level to pollution was Tokyo's "Factory Pollution Prevention Ordinance" in 1949 (Gresser et al. 1981,

16; Hashimoto 1989, 8–9). The ordinance authorized the metropolitan government to limit or even shut down factory operations if public welfare was at danger. Because it lacked teeth, the ordinance proved ineffectual. There were no standards, either for emissions or factory operations, and more to the heart of the ineffectualness, there were no penalties for violations. Even so, the ordinance was imitated by the prefectural governments of Kanagawa (in 1951), Osaka (in 1954), and Fukuoka (in 1955).

The Ministry of Health and Welfare (MHW) was the government agency nominally in charge of environment pollution control in the early postwar period. However, it was not one of the "heavyweight" ministries. It had a small budget and no enforcement powers over pollution violators. Efforts by the MHW to curb pollution abuse were largely suppressed by the more powerful ministries such as the Ministry of International Trade and Industry (MITI), the Ministry of Construction, and the Ministry of Transportation. In particular, MITI considered all regulation (pollution and otherwise) of industry its personal domain and strenuously resisted all efforts by other ministries to undercut that control. MITI's rallying cry, and indeed that of the government in general, was that pollution control would impair economic growth. This ensured that pollution problems were basically ignored by the controlling interests in the national government. The MHW's National Survey on Environmental Pollution illustrates this point.

In 1954 the Bureau of Sanitation of the MHW published the results of a nationwide survey of environmental pollution conducted the previous year (Hashimoto 1989, 14). The survey was based on pollution-related illness complaints filed by citizens with prefectural governments. The survey found that over 10,000,000 complaints had been filed in over 100,000 pollution cases. In the category of air pollution, the worst problems (in order of number of complaints filed) were: (1) dust and particulate matter, (2) offensive odors, (3) gases, and (4) smoke. Based on the results of the survey, the MHW began drafting pollution control legislation in 1955. However, pro-development economic groups such as the Keidanren and various chambers of commerce, and ministries such as MITI and the Ministry of Construction, adamantly opposed the draft legislation. The Ministry of Finance even held up funds for preparation of the draft legislation (Gresser et al. 1981, 17). In the face of overwhelming opposition, the MHW substantially watered it down. But this did not satisfy opponents, and the agency was forced to withdrawal the proposal altogether.

In 1960 Ikeda Hayato became prime minister pledging to double national income within ten years. Industrial development reigned supreme and an era of ultrarapid growth was inaugurated. Japan hosted the Olympics in Tokyo in 1964. This event represented Japan's full reentry into the international political community and its emergence as an economic power to be reckoned with in the global economy. Needless to say, environmental protection was not part of the development agenda, though feeble efforts to bring pollution under control occasionally emerged. One such effort was the 1962 Smoke and Soot Regulation Law.

1962: Smoke and Soot Regulation Law

Japan's first attempt at comprehensive air pollution legislation was the Smoke and Soot Regulation Law of 1962 (*Baien no Haishutsu no Kisei tō ni kansuru Hōritsu*, literally Law Concerning Regulation, etc. of Soot and Smoke Emissions) (Gresser et al. 1981, 264; Hashimoto 1989, 15–16). This law established the basis of modern air quality management in Japan. The drafting of the law basically pitted MITI against the MHW, and the MHW basically lost. In preparation for drafting of the law, the MHW made an extensive study of existing programs in England, the United States, West Germany, and other countries, and proposed strict air pollution regulations. In the end, however, the strict regulations were vastly watered down by MITI. The final law regulated emissions of smoke, soot, dust, and SO_2 generated only by certain sources only in designated air pollution control districts. Also, the law did not apply to mines, power plants, or petrochemical works, which included some of the worst offenders.

The law worked well in containing the visible signs of heavy industry—black, dense billows from industrial chimneys that dumped soot on neighboring areas. But it was not effective against SO_2 emissions. SO_2 was regulated at the "stack outlet." The SO_2 emission standard was set as 0.18% to 0.22% of the exhaust gas volume at maximum stack output. Besides the fact that the standards themselves were not strict, they proved easy to circumvent. A company simply built more and/or larger stacks and emitted the same amount of SO_2 but now from more and/or larger stacks, or, diluted the exhaust gases with air to meet the standards. As an example of the feebleness of the law, the emissions from the worst polluters at Yokkaichi (to be discussed later), which was the site of the worst air pollution in Japan at the time, were only 0.17%, when the strictest standard was 0.18% (Gresser et al. 1981, 471, footnote 237). The MHW attempted to have Yokkaichi and Mishima-Numazu (the site of a planned petrochemical complex larger than Yokkaichi) designated as air pollution control districts under the law, but they were defeated by MITI and industry.

Although the law itself was only partially effective, one of the provisions that proved to be highly effective for future regulation was requiring prefectural governments to monitor air quality and to issue alerts if the pollution levels posed a danger to public health. The issuance of alerts was the first time the government disclosed data on pollution levels to the public, and this helped raise public consciousness of the worsening air quality. For all its failures the law did establish Japan's first emission standards, air monitoring programs, and alert measures for the public. However, air quality during the 1960s became almost intolerable. Traffic police wore gas masks and oxygen stations were installed for pedestrians in big cities. The name that came to symbolize the deadly state of air pollution in Japan began to appear more and more frequently in the press: Yokkaichi.

Yokkaichi

Yokkaichi, opened in 1959, was the site of Japan's largest petroleum refining complex.[3] Monitoring of SO_2 began in 1960 and high concentrations were recorded. By 1961 local medical authorities began reporting a sharp increase in the incidence of asthma, emphysema, bronchitis, and other respiratory ailments, especially among the residents of Isozu, a village near the complex. Citizen protests were lodged with the local government, but they were ignored. In 1963 a second complex began operation in the same area. The situation worsened. In addition to high daily and annual average SO_2 concentrations, low rainfall pHs were also reported. At this time the refineries in the complex were using high sulfur imported oil. Also in 1963 a Yokkaichi Pollution Countermeasures Council (*Yokkaichi-shi Kōgai Taisaku Kyōgikai*) was formed of local politicians and labor unions to assist the residents in their campaign against the companies. Soon the term *Yokkaichi zensoku* (Yokkaichi asthma)[4] became a household word in Japan, and the rallying cry "No more Yokkaichi's" became a stock slogan for antipollution and antidevelopment groups (Hashimoto 1989, 11).

The MHW began to sense a crisis developing and so proposed to MITI that a commission be established to study the situation and make recommendations. In 1963 the MHW and MITI established the Kurokawa Task Force, headed by Kurokawa Mutake, then director-general of MITI's Institute of Industrial Technology (Gresser et al. 1981, 265; Hashimoto 1989, 16–17). After six months of study the Task Force delivered a set of thirteen recommendations. They included: (1) instituting special SO_2 emission standards at Yokkaichi, (2) developing and deploying desulfurization technology, (3) constructing higher stacks (the highest stacks were 60 meters at the time) to avoid downwash of SO_2 from the stacks, (4) providing medical care for victims of Yokkaichi pollution-related illnesses, and (5) establishing air quality and meteorology monitoring stations. Although the intentions and the quality of the investigation were admirable, in the end, MITI softened, altered, or thwarted implementation of most of the recommendations. The one recommendation that was implemented straight out, though, was construction of taller stacks. When completed this succeeded in lowering the local SO_2 peak levels, but also succeeded in spreading the problem over a far wider area.

In 1964 after a three-day smog an asthmatic by the name of Furukawa Yoshio died from respiratory difficulties (Huddle and Reich 1987, 73–74). He became the first widely publicized victim of Yokkaichi air pollution, and his death became a symbol of the problem. In 1965 the city began passing out face masks laced with a chemical to purify the air to over 3,000 children at four elementary schools, with little apparent effect (Iijima 1979, 203). Face masks also became a symbol of Yokkaichi air pollution. In 1967, twelve victims, residents of the heavily polluted village of Isozu, filed suit against six companies within the petrochemical complex. Their trial was extensively covered by the press, and the court eventually upheld their case in 1973.[5] After this, and indicative of the speed with which industry could move when motivated, the pre-1971 level of 100,000 tonnes of SO_2 emissions was reduced to 17,000 by 1975. This was achieved by deploying flue gas desulfurization technology and by using fuels with a lower sulfur content. By 1985 the mortality rate for chronic bronchitis and emphysema reached levels similar to unpolluted areas, and blue skies returned to Yokkaichi; but we are getting ahead of our story.

Third Environmental Era (1967–1973): Domestic Environmental Revolution

The Buildup

1967: Basic Law for Pollution Control

In the postwar decades, citizen rebellion against industrial pollution awoke like a tiger from a long sleep. As in North America and Europe, protest against environmental degradation emerged from bottom-up. Ordinary citizens expressed increasing rage at deteriorating environmental conditions. Citizen protest against the abuses of industry in Japan slowly coalesced around the "big four" pollution cases. These cases represent four instances of three types of pollution-related illness (two water-borne and one air-borne). The cases are as follows: (1) and (2) Kumamoto and Niigata Minamata disease caused by industry-related mercury poisoning; (3) Toyama itai-itai disease caused by mining-related cadmium poisoning; and (4) Yokkaichi asthma, described previously.

By the mid-1960s industrial pollution was a singular focus in political campaigns. In the Diet each house formed a Special Committee for Industrial Pollution Control. This triggered a long and intense debate within the government over the future shape of pollution control policy. The MHW championed strict pollution control legislation in which priority was given to public health over industrial development, and in which industry was to be held strictly responsible for pollution control and liable for pollution-related damage. These proposals met with fierce resistance from industry and the pro-industry ministries. In the end a weakened bill was passed in 1967, the Basic Law for Pollution Control (*Kōgai Taisaku Kihon Hō*).[6]

The law was the first in Japan to establish general principles and objectives for overall environmental policy and to provide a legal mechanism for their implementation. Though strange to most Westerners, the seven-page Basic Law for Pollution Control, like other Japanese basic laws (*kihon hō*) in other areas, contains no specific, legally binding provisions. Instead it broadly sets forth principles, administrative structures, general methods of implementation, responsibilities of various parties (central government, local governments, businesses, and citizens), and a framework for future legislation. Specifically, the 1967 law empowered the central government to set standards for air, water, and noise quality (so-called environmental quality standards), spelled out the mechanisms for drafting pollution-control programs and aiding victims of pollution-related diseases, mandated submission of an annual report to the Diet on the state of the environment, established an Environmental Pollution Countermeasures Council (*Kōgai Taisaku Kaigi*) within the Prime Minister's office, and stressed the desirability for environmental planning (but with no legal obligation requiring submission of environmental impact assessments for large public or private projects, in sharp contrast to the impact assessment requirement included in the U.S. National Environmental Policy Act [NEPA] of 1970).

Article 1 states that the purpose of the law is "the protection of the public's health together with the preservation of the living environment." Living environment was defined as that part of the natural environment "closely related to human life." Thus, nature conservation per se was not an intention of the law. The word *kōgai* ("public injury") was defined as "any situation in which human health and the living environment are damaged by air pollution, water pollution, soil pollution, noise, vibration, ground subsidence and offensive odors that arise over a considerable area as a result of industrial or other human activities." This definition in essence circumscribed the coverage of the law to these specific forms of pollution. Besides the restricted scope of coverage of the law, the main criticisms were the inclusion of a harmonization clause (namely, environmental pollution control policies were to be "harmonized" with economic development), the lack of strict liability of industry for pollution damage, and the lack of an environmental impact assessment requirement.

1968: Air Pollution Control Law

One of the first laws enacted that specifically implemented the guidelines set out by the 1967 Basic Law on Pollution Control was the 1968 Air Pollution Control Law (*Taiki Osen Bōshi Hō*). This law replaced the 1962 Soot and Smoke Regulation Law. Its most noteworthy feature was establishment of an emission standard for SO_x, which replaced the ineffectual earlier standard. The law employed a "k-value" system for emission control.[7] The purpose of the system was to regulate the maximum ground level concentration of SO_x for a designated district based on a calculation of the effective stack height for given sources in the district. There were several problems, though. First, it encouraged many industries to build taller stacks that, while relieving the local problem, spread it to wider areas. Second, the behavior of pollutants did not necessarily conform to the diffusion theory that was the basis of the k-value system. And third, the designation of air pollution districts was more a political than environmental process. The k-value system lasted six years before it was abandoned.

1969: AAQS for SO_2 and MITI's Ten-Year Desulfurization Plan

The first ambient air quality standards (AAQS) were set for SO_2 in 1969 by Cabinet order. The annual average SO_2 standard was 0.03 ppm. To meet the standard, a Ten-Year Desulfurization Plan was instituted by MITI in the same year (Hashimoto 1989, 23–24). The plan called for the installation of desulfurization equipment, increased use of low sulfur coal and oil, and expansion of nuclear power. It represented the first integration of energy, pollution control, and public health policies in Japan. The combination of the Air Pollution Control Law, the SO_2 AAQS, and the Desulfurization Plan inaugurated a downward trend in SO_2 concentrations. However, the fight over pollution control was far from over.

In 1969, the Tokyo city government passed its own stricter-than-the-national-government pollution control ordinance (Gresser et al. 1981, 246–247). Among other provisions, it contained tight emission standards and held industry liable for pollution injury. This instigated a running battle with the central government over pollution control and added fuel to rapidly growing pressure to revise the Basic Law for Pollution Control, which was then only two years old.

The Revolution

In the early 1970s a number of incidents occurred to vastly accelerate the pace of change in environmental pollution control.[8] The central government was still laggard and industry was still antagonistic to pollution control; however, the following events speeded Japan's domestic environmental revolution.

1. A national survey of environmental pollution in rivers and lakes in Japan was published in 1970. The results were not pleasant.

2. In May 1970 newspapers reported incidents of lead poisoning of residents who lived along the roadside in Yanagi-chō in Tokyo. This prompted a major public outcry. MITI almost immediately pledged to cut the lead content of gasoline, and soon implemented a plan to completely eliminate lead from gasoline in five years.

3. On 18 July 1970 the first officially recognized incidence of photochemical smog occurred in Tokyo. More than forty students were reported to have collapsed while playing outside at a high school in the worst-hit ward.

4. On the same day citizen groups orchestrated the first rallies across the country for restoration of "beautiful nature."

5. In October 1970, Ui Jun of Tokyo University began his legendary "open classroom" on environmental and pollution problems. The class was directed at residents and workers, besides students. It was destined to become a popular forum for open and frank discussion of environmental problems and received widespread attention because of its association with Japan's top-ranked university. Ui Jun was subsequently forced out of the university.

6. Episodes occurred of respiratory difficulties and eye and skin irritation resulting from photochemical smog and/or "moist air pollution," particularly in the Kantō Plain around Tokyo. These episodes peaked around 1973, and received widespread news coverage. The "moist air pollution" episodes triggered Japan's first official investigation into the acid rain phenomenon. (This is described in chapter 7.)

7. Between 1971–1973 the verdicts in the "big four" pollution cases were announced. In each case victims had formed groups to bring lawsuits against the companies responsible. Although Japanese legal

requirements made their cases extremely difficult, in all four cases the verdicts emphatically favored the victims and established the right of the victims to compensation. The dates of the decisions were as follows: Niigata Minamata disease (1971), Yokkaichi asthma (1972), Toyama itai-itai disease (1972), and Kumamoto Minamata disease (1973). Publicity surrounding these cases captured the attention of the whole nation.

8. The United States' National Environmental Policy Act (NEPA) was signed in 1970, and the U.S. Environmental Protection Agency (EPA) established. Also, President Nixon indirectly criticized Japan for unfair trade practices by neglecting the environment and not incurring the costs of domestic pollution control.

9. The Organization for Economic Cooperation and Development (OECD) established an Environmental Committee in 1970 and began advocating application of the Polluter Pays Principle (PPP).

10. The 1972 Stockholm Conference on Humans and the Environment brought environmental issues to world attention.

1970: Pollution Diet

In response to the increasing antipollution clamor, Prime Minister Sato established the Headquarters for Pollution Control (*Kōgai Taisaku Hombu*), headed by the Prime Minister himself. In short order the Headquarters drafted fourteen pieces of legislation (an amendment of the Basic Law for Pollution Control, and thirteen anti-pollution laws or amendments), and a special Diet session was called in November and December of 1970. All fourteen pieces of legislation passed (see Hashimoto 1989, 27–28 for a discussion of the air pollution control measures). The "harmonization clause" was deleted from the 1967 Basic Law after heated debate. This clause had been used to justify tilting pollution control measures in favor of industry. The 1968 Air Pollution Law was amended to eliminate the system of designated air pollution districts. Air pollution standards were now applied to the whole of Japan, not just formally designated districts. In addition, a national minimum emission standard was established, and local governments were allowed to set more stringent emission standards than the national government standard.

The Aftermath

1971: Environment Agency

One result of the environmental revolution was the establishment of the Environment Agency of Japan (EA) in 1971.[9] Before its establishment, jurisdiction over environmental problems was distributed among some eleven ministries and almost as many advisory councils. For instance, the 1962 Smoke and Soot Regulation Law was jointly administered by the MHW and MITI, each of which established their own industrial pollution sections to administer the law. Each new pollution crisis was basically met by unilateral, uncoordinated actions by those agencies who claimed the problem area as their territory. After the "Pollution Diet" of 1970 it was clear that there were still serious bottlenecks in the administration of pollution control in Japan. By this time the United States and several European countries had established independent agencies with regulatory powers.

In 1971 the law creating the EA passed without difficulty. The agency was headed by a director-general who was a member of the Cabinet and who held the rank of minister. It took over jurisdiction of the Air Pollution Control Law from the MHW and MITI and established the Air Quality Bureau to administer it. The EA has jurisdiction over not only pollution but also nature conservation issues, has the power to set regulatory standards and monitor pollution levels (which previously were scattered among many agencies), and maintains a research arm (the National Institute for Environmental Studies). The EA was elevated to the Ministry of Environment in January 2001.

1972–1973: Momentum Continues

In 1972 the EA issued vehicle emission standards for the first time.[10] It issued Japan's fourth AAQS in 1973 for NO_x.[11] (The first was for SO_2, the second for CO, and the third for suspended particulate matter.) Also in 1973 the Pollution Damage Compensation Law (*Kōgai Kenkō Higai Hoshō Hō*), the most advanced in the world, was passed. It was heavily influenced by the Yokkaichi air pollution court decision of the previous year.[12]

1973: Oil Shock

The Arab oil embargo of 1973 symbolizes the end of Japan's domestic environmental revolution and a shift of domestic interests to economic and other concerns. The oil shock cooled the central government and industry's ardor for pollution control. The environmental victories gained over the previous years were not lost, though. Japan moved into a consolidation phase.

To summarize, the 1967 Basic Law for Pollution Control ushered in Japan's third environmental era (the domestic revolution, 1967–1973); the era peaked with the emergency session of the National Diet in 1970, the so-called "Pollution Diet;" and ended with the 1973 oil shock. This brief but intense era saw a whirlwind of activity related to the environment. All political actors were in motion simultaneously—victim and citizen groups, central government ministries, industry, local governments, political parties, the National Diet, scientists, the mass media, and the courts. After fierce resistance, industry relented and embarked on a complete overhaul of industrial practices that sparked a revolution in control technologies and monitoring. By the time of the Arab oil embargo the basic foundation for environmental policymaking and administration in Japan was set. Though portions of this infrastructure have changed, it is essentially the same as exists today.

The domestic environmental revolution also saw the establishment of the fundamental set of policies that would eventually become associated with the acid deposition problem in Japan. Although these policies were originally designed to address ambient air quality problems, to the degree that they addressed the emission of acid deposition precursor substances (such as SO_2 and NO_x) they became incorporated into the set of policy options applicable to Japan's acid deposition problem. These included the setting of national standards, establishment of a nationwide monitoring system, use of low sulfur coal and oil, construction of power plants using liquefied natural gas (LNG) or nuclear fuels, deployment of a whole battery of desulfurization and denitrification devices, energy conservation and efficiency measures, strict liability of industry for pollution injury, establishment of a pollution victim compensation fund with a tax levied on industry based on emission quantities, requirement that certain industries have a pollution control manager who is directly responsible for pollution control within a factory, and promulgation of point source and motor vehicle emission standards.

Quiet Acid Deposition Research

The field of air pollution research had a full head of steam by the late 1950s. The Air Pollution Society (*Taiki Osen Gakkai*) of Japan was founded in 1959, and articles on air pollution begin to appear in significant numbers by the mid-1960s. The journal *Taiki Osen Kenkyū* (Air Pollution Research) was founded in 1965. Establishment of the National Air Pollution Monitoring Network (NAPMN) in 1965 spurred further research. This constitutes the backdrop against which a few scattered research efforts on acids in the atmosphere took place.

Kobe 1935–1961 Data Set

The monitoring of precipitation chemistry begun in the 1930s by the Central Meteorological Observatory's Chemical Analysis Section proceeded until 1961, ironically ending just as acidification of the urban atmosphere was becoming recognized as an important issue. During this time, a few local government research institutes also began their own monitoring. As mentioned earlier, the first analysis of one of the Chemical Analysis Section's 1935–1961 data sets (for Kobe) was not published until some twenty-five years after data ceased to be collected (Ishikawa and Hara 1997). The Kobe data set is the oldest and longest in Asia for which the exact measurement techniques are known.

There are older data sets, for instance the Nishi-ga-Hara data set from 1913–1933, but information on the techniques has been lost. Knowledge of measurement techniques enabled Ishikawa and Hara to analyze the quality of the data. The Kobe data (see figure 6.1) shows that the pH was less than 5.5 for the first years (Miyake reported in his 1939 paper a value of 5.2 for Kobe for the years 1936–1938); gradually rose to a value of about 6.0 by 1945; was constant between 1945 and 1952; and then steeply declined to a value of around 4.5, which was maintained until 1961 when measurements ceased. The 1961 value is roughly the same as found in Kobe today. Ishikawa and Hara's analysis revealed the close parallel between industrial activity and historic changes in pH levels in Japan.

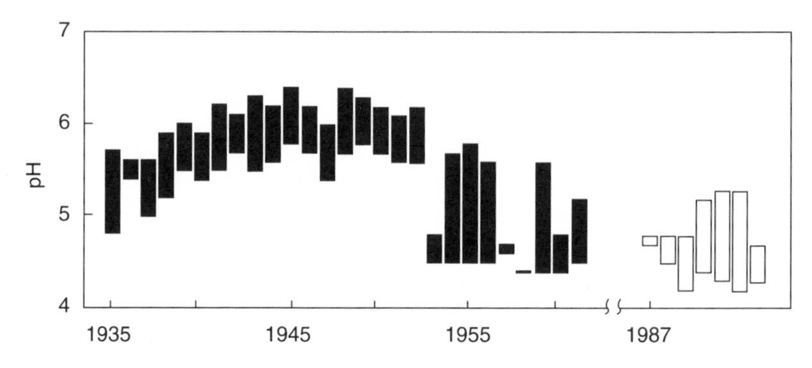

Figure 6.1
Temporal variation of pH at Kobe (1935–1961). Reprinted from Yuriko Ishikawa and Hiroshi Hara, "Historical Change in Precipitation pH at Kobe, Japan: 1935–1961," *Atmospheric Environment* 31, no. 15: p. 2369, figure 3, 1997. Used with permission from Elsevier.

1964: Deposition Analysis in Tokyo

Despite the fact that air pollution research skirted the problem of acidification of the atmosphere, the collection of bulk dry and wet deposition to monitor pollutant fallout from the atmosphere (especially dust and soot) in the 1960s was a back door through which interest in acidification problems entered. The researcher who was destined to become the leading figure in early acid deposition research got his start on the problem doing dustfall monitoring. In the 1960s Okita Toshiichi of the Ministry of Health and Welfare's National Institute of Public Health in Tokyo began publishing articles on acidic substances in the atmosphere (primarily, sulfate) gathered in bulk deposition sampling. Eventually he moved to the University of Hokkaidō in Sapporo, then returned to the Kantō area to become head of the Atmospheric Environment Section of the National Institute for Environmental Studies in Tsukuba, and finally transferred to Obirin University near Tsukuba, from which he retired in 1997.[13]

In 1962–1963 Okita and a collaborator performed an analysis of the deposition of various cations and anions in the Tokyo area (Okita and Konno 1964). The purpose of their research was to determine the spatial distribution of air pollutants in Tokyo. Since at this time there was as yet little observational data, and no network to measure pollutant distributions, they attempted to infer them from an analysis of the distribution of deposition rates of various chemical species in rainfall. This was the first analysis of this type in Japan. Over a one-year period between September 1962 and August 1963 at twenty-eight locations throughout the Tokyo metropolitan area they measured cation (Ca^{2+}, Mg^{2+}, Na^+, K^+, NH_4^+) and anion (SO_4^{2-}, NO_3^-, Cl^-) deposition rates.

Okita and Konno divided Tokyo into four areas: residential, industrial, business, and mixed business-residential. In all cases (with one slight exception) they found that the annual average deposition rates of all species were smallest in residential areas, the areas furthest from the city center and industrial belt. Comparing the deposition rates in the other two areas to that of the residential area, they found that the ratios, while greater than one, were still close to one (i.e., the deposition rates in all areas were roughly the same) for all species except SO_4^{2-} and Ca^{2+}. For these two species the deposition rates in the city/industrial centers were 2 to 2.5 times larger. They speculated that the high Ca^{2+} deposition in these areas was due to the lack of road pavement, which resulted in soil particles being blown into the air by wind and traffic activity, and that the high SO_4^{2-} was due to coal combustion.

Okita and Konno also made a comparison of deposition rates and rainfall amount. They were able to differentiate dry and wet deposition rates because for three months of the one-year measurement period rainfall was extremely small (and was zero for one of these months). They assumed pollutants during this three-month period were dry deposited. They concluded that dry deposition was dominant for SO_4^{2-}, NO_3^-, and Cl^-. Measurements of pH were not included in the research. It is interesting to note that their paper contains no references to previous Japanese works. As a matter of fact, the paper contains no references at all, though there is mention of suspended particulate matter measurements in Frankfurt and Berlin in Germany.

1965–1972: Okita's Sulfate Research

Not long after the previous research was published Okita wrote several articles on the conversion of sulfur dioxide to sulfate in the atmosphere (Okita 1965, 1967, 1968). He makes clear he is following the lead of European and American researchers who began publishing on the topic in the 1950s. In all three papers, Okita's research is set in no clear environmental context other than a loose connection with air pollution. Later, Okita (1972) returns to the $SO_2 \Rightarrow SO_4$ conversion problem, but now armed with the knowledge that the problem is one of international importance. He describes the debate raging in Sweden over the ecological impacts of acidic precipitation, and references the first issue of *Ambio*, a journal published by the Royal Swedish Academy of Sciences starting in 1972. He comments that low pH rainfall and ecological impacts were currently observed in Japan (see the discussion that follows on the Yokkaichi and Kumamoto pH data sets). A manifest shift in emphasis can be detected in this paper, a shift from what we might call an "SO_2 air pollution" orientation to a broader "acid rain" orientation. The shift was clearly prompted by an awareness of the emerging acid deposition issue in Europe.

1970: Bleached Morning Glories

A brief announcement (Fujii 1971) appeared in the magazine *Air Pollution News* describing the observation of bleached morning glory flowers in association with acidic rain in the Kinki (Osaka-Kyoto) district of Japan in the summer of 1970. In response to this announcement, eight years of pH data from Kumamoto in Kyūshū in southern Japan was published in a later issue of the same newsletter (Nishi 1971). In the article Nishi Teizō notes that there seems to be a connection between air pollution and low pH in rainfall. In his analysis of Kumamoto data he demonstrates that particulate matter concentrations drawn by a high volume sampler drop drastically when it rains. Thus, he concludes that air pollutants are being washed out by rain. He goes on to present pH data from 1963 to 1970 that shows a decline in the annual average pH from 7.0 in 1963 to 5.8 in 1970. This result he finds puzzling because pHs were decreasing even though Kumamoto had no large polluting industrial plants. He speculates that changing fuel consumption patterns, an increase in automobiles and trucks, or refuse incineration, may be the cause. He also notes that the pollutants may be originating from industry in the towns surrounding Kumamoto. He concludes that it is urgent to seek the causes of the increasing acidity of the rain.

1971: Yokkaichi pH Measurements

In response to the above two announcements, Yoshida Katsumi (1971) published data for Yokkaichi for the years 1961–1967 supporting the same acidification trend. During this time, pH dropped precipitously from about 6.3 to 4.0, more than a 100-fold increase in acidity. Along with the short one-page announcement is a picture taken at Yokkaichi on an early August morning in 1970 after a light rain, showing a poor morning glory flower speckled with burn spots caused by acids in the rain.

1971: Takeuchi and Do-It-Yourself Sulfate Analysis

In 1971 an interesting piece of "homemade" sulfate research was published, indicative of the somewhat primitive and self-initiated nature of research on acidic pollutants in Japan at the time. An elementary school teacher, Takeuchi Ushio, published his "do-it-yourself" analysis of the sulfate content in rainwater conducted between 1967 and 1970 at the location of his elementary school in Tokyo (Takeuchi 1971). The experimental setup is noteworthy for its simplicity. A one-meter square piece of vinyl with a hole in the center was stretched over a wooden frame 40 centimeters off the ground, and rainwater captured in a bottle placed beneath the hole. The sulfate content was measured home-chemistry-set style.

From this simple method the following results were derived. First, no clear pattern was discovered in the monthly averages over the four-year period. There were high and low values in both summer and winter. Takeuchi contrasts this with the pattern Miyake found in 1939 in Tokyo of a winter high and summer low sulfate concentration. Second, Takeuchi compares his annual average values to the few earlier measurements made in Japan. These are shown in table 6.1. (Note: There is some uncertainty in the table as to the year and place because the sources of the data are not well referenced in the article.) The table indicates in a rough way that the sulfate concentration in rainwater had increased dramatically since the end of the war. In Nagoya it increased some thirteen-fold. Takeuchi's four-year results also suggested an increasing sulfate trend.

Table 6.1
Takeuchi's historical comparison of annual average SO_4^{2-} concentrations (1946–1970)

Year	Place	Researcher	Annual average SO_4^{2-} (ppm)
1946/47	Nagoya	Sugawara	0.34
1964/65 (?)	Nagoya (?)	Sugawara et al.	4.5
(?)	Tokyo	Miyake	12.0
1967	Tokyo	Takeuchi	4.7
1968	Tokyo	Takeuchi	5.3
1969	Tokyo	Takeuchi	6.5
1970	Tokyo	Takeuchi	8.0

Source: Takeuchi (1971, 396).

A third trend, which Takeuchi found when analyzing SO_4^{2-} concentration as a function of wind direction, was that when the wind blew from the industrial area of Tokyo toward his elementary school the sulfate concentration was double or more compared to when the wind blew from any other direction. Takeuchi later teamed up with a Nagoya University professor to publish a much more sophisticated analysis of sulfate concentrations in precipitation in Tokyo (Nakai and Takeuchi 1974).

1971: World's First Measurement of Volcanic SO_2 Emissions

Until the 1970s, research on SO_2 emissions focused primarily on measuring emissions from anthropogenic sources, such as coal-fired power plants, oil refineries, ore smelters, and pulp and paper mills. In large part due to the difficulties of measurement, emissions from natural sources such as volcanoes, forest fires, and peat bogs were unknown. In 1971 the previously mentioned Okita used a correlation spectrometer originally developed for remote measurement of air pollutant fluxes from industrial stacks by a Canadian firm, Barringer Research of Toronto, to measure SO_2 emissions from an active volcano. He was the first person in the world to directly measure volcanic emissions. In April of that year Okita measured the rate of SO_2 emission from Mt. Mihara, an active volcano on Oshima Island in the Pacific Ocean about 130 kilometers south of Tokyo.

When Okita first obtained the instrument he measured SO_2 and NO_2 emissions in Japan's industrial heartland, the Kawasaki district south of Tokyo (it was for this type of measurement that the instrument was originally designed). Then, because of his long interest in volcanoes, he decided to measure volcanic emissions.[14] The results were published in Japanese in two obscure locations (Okita 1971a, 1971b).

The second measurement of volcanic emissions also took place in Japan. However, Okita was not involved. Environmental Measurements, Inc. of San Francisco, California, another licensee of the Barringer technology, upon hearing of Okita's results sent a team to Japan to conduct a survey of emissions from Mt. Asama in June of 1972 (AirNote no date, 1972(?)). The third, and best known, early publication on volcanic emissions is Stoiber and Jepsen (1973). They—one of the authors was from Environmental Measurements, Inc.—measured, using the same Barringer correlation spectrometer, SO_2 emissions from volcanoes in central America in the fall of 1972 and published their results in *Science* in 1973. What distinguishes Stoiber and Jepsen's results from Okita's is not only that they were published in English in a prestigious journal, but also that they were used to extrapolate a rough estimate of the global annual SO_2 emissions from volcanoes (which they estimated to be one-tenth of global anthropogenic emissions). They also compared their estimate with various other estimates based on global-scale geochemical considerations.

The story of Okita's SO_2 measurements, even though it did not address the conversion to acidic compounds, was told here because it illustrates several points about science in Japan that are relevant to both historic and current research on acid deposition. The first point is that because work by Japanese scientists is often published in the difficult-to-master Japanese language, much worthy research in Japan has been, and still is, overlooked by the rest of the world. Okita's situation is a case in point. Even as late as 1994, by which time remote measurement of volcanic SO_2 emissions was commonplace, Okita's initial efforts were overlooked. (See, for instance, Caltabiano et al. (1994) for an example of this oversight.) The second point is that Japanese researchers often failed to make the creative leap from immediate results to new and extended applications, interpretations, comparisons, hypotheses, etc. Again,

Okita's results are a case in point. He failed to put his results into a wider global geochemical perspective that would have further drawn attention to their merit. Okita published one final paper (1975) on volcanic emissions before he was swept up by acid rain research. In this paper, on volcanic emissions from Mt. Asama after an eruption in 1974, he now situates his results in the context set by Stoiber and Jepsen, and reestimates the global discharge of SO_2, increasing it by a factor of three.

1973: Air Pollution and Forest Decline in the Tokyo Area

Yambe Yoshito of the Forestry and Forest Products Research Institute published the first extensive study of forest decline related to air pollution in Japan (Yambe 1973). The institute started a research program on "The Impact of Air Pollution on Trees" in 1965. Based primarily on comparative analysis of photographs taken starting in 1966, Yambe showed that by 1973 a noticeable decline in tree health (especially in industrial areas and along heavily used roadways) had taken place in the Kantō Plain surrounding Tokyo. He correlated the level of decline with the distribution of SO_2 ambient air concentrations.

Yambe's research was considerably refined in a second article (Yambe 1978). Based on the above-mentioned photos, a five-level classification scheme for *sugi* (Japanese cedar) and *keyaki* (zelkova) was devised—from healthy to dying (where dying was defined as wilted crown, high loss of needles/leaves, and numerous dead branches). Then for three years from 1972–1974 tree health in a 100-kilometer radius from the center of Tokyo was carefully surveyed. Also, in 1975 a survey was conducted along the coast of neighboring Chiba Prefecture, and in 1976 a survey was conducted in Kyoto. The results of the Tokyo survey showed that the area of dying trees extended some 25 kilometers from the city center. Tree health improved the further one traveled from the city center. Trees were fully healthy by the time one reached the foothills of the mountains 50 to 100 kilometers away. None of the other urban areas surveyed showed tree decline as severe as Tokyo. The physical-chemical mechanisms of tree decline were not investigated. Neither paper mentions acid rain as a probable cause. As we shall see, it was not until the mid-1980s that acid rain appeared on the list of possible causes.

Analysis of Period 3

By 1974, on the eve of the dramatic entrance of Japan's first headline-catching acid rain episode, there were but few scientists (and fewer if any policymakers) interested in acids-in-precipitation. Background scientific knowledge such as accumulated during the active 1920–1945 period was still being quietly added to, but an integration of the sort occurring in Europe in the postwar period was not forthcoming in Japan. Acid deposition science silently rode the wave of air pollution research.

An epistemic shift from an incipient *acidified urban precipitation* problem-framework to a vague but generalized *acidified atmosphere* framework distinguishes Japan's second acid deposition period from its third. Researchers at local institutes contributed significantly to the expanded framework. Thus, the shift was prompted not only by continued importation of Western scientific ideas related to atmospheric chemistry, but also widespread dissemination of precipitation chemistry techniques within Japan.

As this shift was taking place in Japan, in Europe, beginning in the 1950s and centered in Scandinavia, researchers were homing in on an ominous new environmental problem of international dimension—a problem that came to be known as acid rain. Toward the end of Japan's third acid deposition period the acid rain problem burst onto the international political scene in Europe. Thereafter, its associated scientific problem-framework slowly and fitfully migrated to Japan. It is to the new acid rain problem-framework that was being created in Europe that we turn first.

Acid Deposition in Europe and North America

The first precipitation chemistry network in Europe was started by Hans Egnér, a Swedish soil scientist, who began systematic observations during World War II of the relationship between nutrients in the atmosphere and plant growth.[15] After the war in 1948 he organized this into a nationwide network in Sweden. Sampling buckets were set out at experimental farms throughout Sweden and all that fell into the bucket (dry and wet deposition) over a one-month period was analyzed. One of the quantities measured was pH. This network eventually grow from Sweden to include Norway, Denmark, and Finland, and constituted the first large-scale precipitation chemistry network in Europe and the world. In 1956 the Stockholm International Meteorological Institute took it over, and named it the European Atmospheric Chemistry Network. By 1957 it extended to Poland and the USSR, and included bulk samplers at 100 observation sites in northern and western Europe. Monthly samples were collected and analyzed for the concentrations of eleven major ions. The establishment of this network marked the beginning of regional-scale collection of data on the chemistry of European precipitation, which allowed among other things, for analysis of regional-scale changes in precipitation pH.

The first regional precipitation chemistry monitoring network in the United States was established between 1953 and 1955 by H. V. Jordan and his colleagues at monitoring sites set up at State Agricultural Experiment Stations in the southern part of the country. Jordan's efforts were not, however, the first studies of precipitation chemistry in the United States. The earliest data for which pH determination is possible is that of W. H. MacIntyre and I. B. Young who from 1917 to 1922 made an analysis of crop nutrients contained in precipitation in Tennessee. The earliest measurement of precipitation pH in the United States seems to have been made by H. G. Houghton of MIT in 1939 when he measured the pH of a single rainstorm in Maine. For a fuller discussion of the networks and early precipitation chemistry measurements in the United States, see Cogbill (1976) and Likens and Butler (1981).

Inspired in part by the data from the European Atmospheric Chemistry Network, Svante Odén, a Swedish soil scientist, in 1967 locked the many pieces of an environmental puzzle together to reveal an integrated picture of a long-range, transboundary, multiple-impact acid rain problem. Odén's (1967) novel ideas first appeared in a Swedish government report, and a newspaper article (*Dagens Nyheter*, 24 October 1967) that graphically described an insidious "chemical war"—a war being conducted with air pollutants. Soon after, in a comprehensive scientific paper, Odén (1968) outlined in remarkably accurate detail the mechanisms and impacts of acid deposition, and he almost single-handedly created the acid rain problem-framework. He synthesized a wide range of ideas and data into the following hypotheses:

1. Acidified precipitation due to industrial activities was a regional scale phenomenon in Europe. It centered in the Benelux (Belgium-Netherlands-Luxembourg) area, and included much of northern and western Europe.
2. Acidified precipitation in Scandinavia was attributable to long-range transport of sulfur and nitrogen compounds emitted in England, Germany, and central Europe.
3. Precipitation and surface waters were becoming increasingly acidified over time.
4. Acid inputs to soils acidified the soils, displacing nutrient cations, reducing biological nitrogen fixation, and releasing heavy metals (especially mercury) into surface waters.
5. Declining fish populations, decreasing forest growth, increasing plant diseases, and accelerated degradation of materials were all in part due to acidified precipitation.

These hypotheses—stripped of their European context—constitute the first distillation of what eventually became the scientific knowledge core of today's "standard acid deposition problem-framework." The publication of Odén's synthesis sparked a storm of scientific and political activity, and marked the passage of acidification of the environment from a curious topic of scientific inquiry to a controversial topic of political debate. From this point on the science and politics of the now acid rain *issue* became deeply intertwined.

The Swedish and Norwegian governments almost immediately seized on the issue and began a push for political solution. Although unknowns and uncertainties abounded, the "impact message" in the emerging European problem-framework—a general pattern of cause (sulfur dioxide emissions emanating from continental Europe) and effect (negative impacts on Scandinavian fish and forest resources)—was both sufficiently backed by scientific evidence and sufficiently alarming for these two countries to launch a drive to formulate international policy on the issue at the 1972 Stockholm Conference on Humans and the Environment. Sweden presented a case study report at the conference—*Air Pollution across National Boundaries: The Impact on the Environment of Sulfur in Air and Precipitation* (Bolin 1971). This report indelibly placed the acid rain issue on the international political landscape.

North America was not far behind Europe. The man who first discerned a pattern of large-scale, and increasing, acidity in the precipitation of eastern North America was Gene Likens. Although there were others before him, such as Eville Gorham and Harold Harvey in Canada, who investigated acidification of precipitation as a local problem, it was Likens who first sounded the alarm that a regional-scale problem existed. Studying ecosystem dynamics at the Hubbard Brook Experimental Forest in New Hampshire in the 1960s, Likens and his colleagues tracked everything that went into, was cycled within, and came out of the Hubbard Brook watershed. One of the items that went into the system was precipitation and "large" quantities of acids carried in precipitation.

It wasn't until Likens went to Sweden in the late 1960s and talked with Svante Odén and others that he realized that what he observed at Hubbard Brook was quite likely the tip of an acid rain iceberg; that the Hubbard Brook measurements were really pointing to a large-scale acid rain problem like that recently discovered in Europe. Likens and his colleagues then gathered records of previous precipitation chemistry measurements made in the United States, and upon collating and analyzing them became convinced they had discovered the world's second large-scale acid rain problem. Gene Likens, F. Herbert Bormann, and Noye M. Johnson (1972) published the first article warning of the acid rain danger in North America. They argued that acid rain was potentially an international problem in North America, as it was in Europe, and urged "consideration of [their] data in the establishment of air pollution standards and a massive effort to increase our understanding of this problem" (Likens et al. 1972, 40). This article heralded the entrance of acid deposition as a regional-scale environmental problem and international political issue in North America.

Japan lay completely outside the mainstream of this international activity on acid rain. By the early 1970s Japanese researchers, and at the 1972 Stockholm Conference, Japanese policymakers, were made aware of the existence of the issue and this began to influence their thinking. But Japan remained preoccupied with domestic air pollution, not regional-scale acid deposition.

Acid Rain Hovers outside Politics

Similar to the second acid deposition period, Japanese scientists vaguely perceived an acid deposition problem during the third period, but vague problem recognition did not result in construction of a clear-cut problem-framework. However, Japanese researchers again, as in the second period, had its basic components. They retained a rough idea of cause-and-effect (e.g., Okita's and Takeuchi's sulfate analyses for Tokyo); they had a handle on the extent of the problem (e.g., pH data, albeit uncoordinated, from local research institutes on a wider scale than in the second period); and they had a sense of impacts (e.g., Yambe's research on trees in the Tokyo area). Once again, though, the pieces of the puzzle were not put together.

With no problem-framework, acidification of the atmosphere only tangentially entered the intense domestic political debate on air pollution. The lack of a well-defined problem-framework inhibited formation of science-to-policy bridging objects. Japanese scientists continued to admirably utilize the 'universal' science-to-science bridging objects when adapting Western acid deposition-related knowledge to the Japanese context, but the lack of a domestic problem-framework prevented development of bridging objects that might have influenced the policy process. Thus, there were no cause-and-effect, extent-and-intensity, or impacts bridging objects to allow political actors to calculate their interests or to foster public dialogue and political action. The lack of a relationship between acid deposition science and policy during the third acid deposition period is illustrated in the spare diagram in figure 6.2.

Pollution-related precipitation chemistry methods and knowledge continued to be imported. Once imported, localized knowledge and methods were refined (e.g., Okita's sulfur dioxide to sulfate conversion work) and disseminated within the country. The knowledge base expanded beyond the urban atmosphere to acidification of the atmosphere in general (e.g., volcanic emissions). What stands out during the third period is the quiet but diligent efforts of a handful of individuals (about two dozen). These individuals constituted a small but stable, and second generation, expert community. Information was shared between researchers and broadcast through journal articles and presented at conferences, but the exchange was sporadic and uncoordinated. No fora existed for focused dialogue

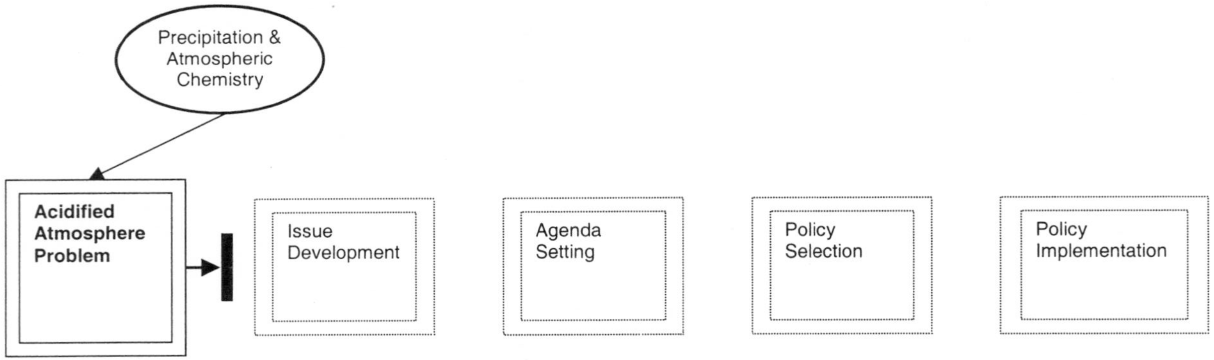

Figure 6.2
Science-policy interface of the acidified atmosphere problem (1945–1974)

on acidification matters. Even so, the historical scientific momentum on the acid deposition problem continued to build.

Japan's scientific foundation was such that it was positioned to absorb the newly created European acid rain problem-framework. Toward the end of the period, this problem-framework began filtering into Japan. In Okita's writings we can clearly follow the evolution in his awareness of atmospheric acidification problems. His first papers lack an environmental context and have an isolated, detached quality to them. They contain few references to foreign work. But by the early 1970s there appears an almost delight in being able to situate his research within the context of an important international environmental problem. However, Japanese scientists were not ready to act upon the new framework. Indeed, they soon found themselves drawn to an acidification phenomenon quite different from the one addressed in the European acid rain problem-framework. A decade-long hiatus resulted before Japan was receptive to full importation of the Western framework.

Sustainability

During its third acid deposition period, Japan again experienced a sustainability crisis. Industrial pollution-induced illnesses became a paramount issue in domestic politics. In contrast to Japan's first crisis at the turn of the century, however, acid deposition was a minor factor in the debate. For this reason, the postwar crisis was not discussed in detail in this chapter. Although acid deposition was not central to the larger sustainability debate in Japan in the 1960s and 1970s, several key developments for eventually addressing acid deposition sustainability took place during the third period, as follows: (1) continued acid deposition-related scientific capacity building and accumulation of scientific knowledge, (2) the emerging role of "environmental scientists," and (3) the beginnings of an epistemic shift to the new Western acid rain problem-framework.

Similar to the second period, the process of adding to acid deposition-related knowledge, and the arduous task of building domestic scientific capacity continued, even while industry was careening toward a sustainability breaking point. Research capacity spread to local research institutes, and they too began to add detail to the storehouse of Earth space/time-specific knowledge on atmospheric acidification in Japan. National-level researchers remained the scientific leaders, though. Without a solid problem-framework, researchers did not get off the ground developing a solution path for atmospheric acidification. However, awareness began growing among scientists of this period that developing such paths (as were developed in the cases of Minamata, itai-itai, and Yokkaichi illnesses) was part of their mandate as environmental scientists.

Thus, over and above adding to scientific knowledge, the role of scientists in environmental politics was changing in Japan by the 1960s. As a result of massive research efforts related to industrial pollution, they were becoming skilled in elucidating cause-and-effect mechanisms and in significantly shaping policy outcomes.[16] In addition, in the Yokkaichi case, we find that besides elucidating the cause-and-effect relationships between air pollution and respiratory illness, scientists engaged in the first large-scale, systematic environmental impact assessment in Japan (the Kurokawa Commission). From this time on there emerges what we might call the first true environmental scientists in Japan who proactively engage in outlining solutions to environmental problems, another cornerstone in developing a sustainable society.

7

Period 4 (1974–1983): Moist Air Pollution

Fourth Environmental Era (1973–1990): Transition

Japan's fourth environmental era (1973–1990)—the transition era—spans two acid deposition periods: moist air pollution (1974–1983), and nationwide, ecological acid deposition research (1983–1990).[1] It is characterized by multiple and conflicting trends. One trend was consolidation of the gains made during the brief but intense domestic environmental revolution. The head of steam buildup over the previous decades continued to propel environmental protection activity. The aftershocks of the verdicts in the pollution trials were felt for years. Private investment in pollution control equipment peaked in 1975, reaching almost 18% of total investment in Japan, a figure unmatched anywhere in the world (92 NGO Forum Japan 1992, 78). Another trend was backpedaling on certain aspects of environmental policy, especially in the wake of the two oil embargoes in 1973 and 1979. This regression was most evident in pressure by industry to relax air quality or emission standards, and to weaken the pollution victim compensation system. The upper levels of government and industry began to treat environmental problems as solved. Large-scale and long-stalled public works projects, such as the Seto Inland Sea Bridge, Kansai International Airport built on an artificial island, and extensions for the Shinkansen bullet train were recommenced.[2] A third trend was increased interest in international environmental issues. This is discussed in detail in chapter 9.

Between 1973 and 1990 the structure of Japanese industry began to shift—from heavy manufacturing to the high-tech and service sectors—and many industries began to move abroad taking both jobs and pollution with them. People's lifestyles also began to change. There was more leisure, more emphasis on consumption, and more urban dwellers. Private automobile possession began exponential growth. Resort construction, fueled by the 1987 Resort Law, skyrocketed to the detriment of natural areas. After 1973 the central government together with industry dominated environmental politics. The victim and citizen groups that had been the driving force behind environmental policy change in the 1950s, 1960s, and 1970s lost much of their momentum.[3] It became increasingly obvious that many local groups were bees devoted to their own hive. They formed around specific local issues and that was where their loyalty remained. There was very little in the way of mergers into larger coalitions revolving around larger ecological issues. Very few broad spectrum environmental NGOs developed. Some of the more famous environmental NGOs of the West, such as Greenpeace and the World Wildlife Federation, developed branches in Japan at this time, but their membership remained small and they were all but ignored by the government. As acute pollution problems abated, public interest shifted to amenity issues such as neighborhood greenery, recreation areas, and so on.

Local governments remained a strong actor. The courts more or less receded into the background after the "big four" pollution trials, but on occasion still rendered decisions that carried national impact. The mass media remained a strong voice for environmental protection, but did not retain the forefront role that it had assumed in the 1960s and early 1970s. The role of environmental science and scientists continued to evolve. The National Institute for Environmental Studies was established in 1974,[4] and scientific data was routinely used to inform policy decisions. Science became prominent in the ecopolitical landscape as attention shifted from blatant pollution episodes to more subtle, long-term environmental problems such as global climate change, stratospheric ozone depletion, and regional-scale acid deposition.

A period of consolidation of domestic environmental policies began in the 1980s. The second oil shock of 1979 further shifted Japan's industrial attention away from pollution to energy efficiency concerns. By the end of the decade Japan was experiencing an economic boom that lasted from 1987 to 1991 (the so-called "bubble economy") in which, for instance, land prices in Tokyo quadrupled. Little in the way of new and innovative environmental legislation was introduced during the decade of the 1980s.

During the entire fourth environmental era, there were few notable advances in air pollution policy. The policy framework established in the early 1970s persisted throughout. One exception was a system known as "total mass emission control," which was legislated in a 1974 amendment to the 1968 Air Pollution Control Law.[5] This legislation was designed to remedy the defects of the k-value system. Total mass emission control places a cap on the total emissions in a designated area (thus, it is also often referred to as the "area-wide total pollutant load control system"). Japan was the first nation to institute this form of pollution control. It was put into effect for both air and water pollution. At first SO_2 was the only air pollutant designated, but because of the system's success NO_x was added in 1981. Pollutants and areas to be controlled are stipulated by Cabinet Order. The prefectural governor of a designated area draws up a reduction plan that includes emission and fuel standards for sources in the area. Permits for new industries are granted only if the emission ceiling is not surpassed. Initially, eleven regions were designated for SO_2, but by 1992 there were twenty-four, which accounted for about 30% of both Japan's total population and sulfur oxide emissions, and three regions were designated for NO_x—Tokyo, Yokohama, and Osaka by the EA (EA 1992, 449–452).

As an example of the backsliding that occurred in the air pollution field, the NO_x standard was rolled back in 1978 from the twenty-four-hour 0.02 ppm benchmark to "within the range of 0.06 to 0.04 ppm." In one fell swoop Japan went from 90% of the designated air pollution regions in the country exceeding the standard to 90% being below the standard (92 NGO Forum Japan 1992, 19). This was followed by claims that Japan had now cleaned its air of all major pollutants (smoke and soot, particulate matter, SO_2, and NO_x). It was amidst the progress and backsliding in air pollution control during Japan's environmental "transition" era that acid rain first emerged as a political issue.

Moist Air Pollution

The Trigger Event

Over a two-day period (3 and 4 July) during the summer rainy season of 1974, a staggering—more than 32,000—number of cases of eye, throat, nose, and skin irritation were reported over a widespread area in the Kantō Plain north of Tokyo. There was also damage to crops such as cucumber, eggplant, peanuts, sweet potato, and green beans. The lowest pH value in the area was estimated to be 2.85. This sudden event prompted an immediate response by the national government. The EA formed an investigation committee and launched a five-year project to discover the cause. To understand the prompt political response, one must keep in mind the environmental context of the time. As explained in the previous chapter, this event occurred toward the end of the most intense period of environmental policymaking in Japan's history. The sudden appearance of a massive number of cases of a brand-new medical problem virtually demanded immediate action.

Investigation revealed that this was not the first manifestation of such a problem. The first recorded event had occurred the year before on 28 and 29 June 1973 when 441 cases of eye and skin irritation were reported in an area west and southwest of Tokyo. The pH of this event was estimated to be in the 2.0 to 3.5 range. Because these events were associated with very low pHs, it was at first assumed that the cause of the medical complaints and crop damage was acidic rain. But further gathering of evidence led to the conclusion that the cause could not unequivocally be attributed to acidic rain; therefore, the problem was labeled "moist air pollution" (*shissei taiki osen*). A third outbreak of medical complaints occurred on 25 June 1975, on the first day of the first year of the government investigation (explained later), when 144 cases were reported again in the area north of Tokyo. A pH of 3.05 was recorded at Kumagaya, 100 kilometers to the north of Tokyo.

Thus, acid rain as a clearly recognized public problem first emerged in Japan as a health issue. In total over 33,000 people (mostly middle and high school students) reported to clinics with mild to serious cases of eye, throat, nose, and skin irritation during each rainy season (end of June to the middle of July) over this three-year period. Also, significant crop damage was reported in the same areas at the same time. The three incidents mark the beginning of government and public attention to the acid deposition phenomenon in Japan. They also reveal that the origin of Japanese interest in acid deposition followed a significantly different path than in Europe or North America, at least initially. In both Europe and North America, ecological impacts (particularly aquatic and forest impacts) triggered the first attention. Human health impacts played a negligible role. In Japan the reverse was true, it was human health impacts (albeit on a local scale), not ecological impacts, that first imprinted the acid deposition issue on public consciousness.

Moist Air Pollution Investigation

Starting in 1975 a Moist Air Pollution Investigation Committee (*Shissei Taiki Osen Kentō Iinkai*) set up by the EA conducted research on the moist air pollution problem over a two-week period during each rainy season for five years in the Kantō area. The committee consisted of eighteen members—researchers from national research labs (six), Kantō prefectural research labs (nine), universities (two), and foundations (one). Okita Toshiichi, then head of the Local Environment and Public Health Section of the National Institute for Public Health, headed the committee.[6] Because of its historic importance as the first large-scale investigation of acid deposition in Japan, the committee's work is discussed at length.

The first formal document related to the investigation was a 1975 hand-written report (Moist Air Pollution Investigation Committee 1975) titled "*Shissei Taiki Osen (Sansei Kōu) nitsuite*," which translates as "About Moist Air Pollution (Acid Precipitation)." This is Japan's first official acid rain document. It contains a detailed compilation of available data for the 1973 irritation complaints in Shizuoka and Yamanashi prefectures (the first event), and the 1974 incident in the Kantō area (the second event), including the number and distribution of medical complaints, the air pollutant concentrations, pH measurements, and weather conditions at the time. It then reviews precipitation chemistry data so far collected by prefectural environmental research institutes in the Kantō area. It also references the 1967–1973 Nakai and Takeuchi SO_4^{2-} work, unpublished pH data from the Tokyo

Meteorological Research Institute of the Meteorology Agency, and pH measurements from Yokkaichi (1961–1966), Kumamoto (1964–1972), and Osaka (1970–1972).

Following this section on domestic experiences, the report contains a brief section on foreign experiences. In this case, "foreign" means Swedish and Norwegian experiences. The report discusses research reported in articles in *Ambio* (the journal published by the Swedish Royal Academy of Sciences starting in 1972 only a few years before the moist air pollution investigation), research contained in a lengthy 1974 report published by the Norwegian government, and research then being conducted by the OECD on long-range transport of air pollutants. Thus, the Moist Air Pollution Investigation Committee was clearly aware of acid deposition-related work in Europe.[7]

The investigation itself was divided into five main areas: (1) precipitation chemistry, (2) gas and aerosol ambient air concentrations, (3) meteorological conditions, (4) formation of secondary pollutants and their incorporation into water droplets, and (5) medical experiments on the cause of the human health impacts. Each area is discussed briefly. Results were published in annual reports and a final summary report (Moist Air Pollution Investigation Committee 1976, 1977, 1978, 1979; Photochemical Formation of Secondary Pollutants Investigation Committee—Moist Air Pollutants Subcommittee 1981).

The backbone of the investigation was analysis of precipitation chemistry in the Kantō area. The medical complaints occurred during times of mist or misty rain. Therefore, to understand the problem it was necessary to directly sample precipitation from mist and misty rain. Researchers found these precipitation events difficult to analyze, so a special sampler was invented. The sampler captured each of the first five 100 milliliters of precipitation in five separate glass bottles. In this manner very light precipitation events could be analyzed. This so-called "*yōryōbetsu bunsai hō*" (separate volume-based collection method) was used by each of the seven districts in the Kantō area involved in the study (the prefectures of Ibaraki, Tochigi, Gunma, Saitama, Chiba, and Kanagawa, plus the Tokyo Metropolitan Area). Each district set up one sampler in an urban area and one in a rural area. The number of sampling stations was roughly constant (about fifteen) for the first four years of the investigation, but increased to thirty-three the last year to try to better determine distribution patterns. The samplers operated for ten days in June and July of each year. In addition to analysis of precipitation chemistry in the Kantō Plain area, water samples from fog and mist on three mountains bordering the plain were also analyzed.

Gas and aerosol concentrations were measured at the same stations where precipitation samplers were located. Scientists made two attempts to construct models of acidification processes in the atmosphere. Both focused on the formation of secondary pollutants and their incorporation into water droplets. The first was a cloud water acidification model based on field measurements, and the second was a model based on chamber experiments in the laboratory. The cloud water model (Ohta et al. 1981) was only partly successful in reproducing the lowest pH values recorded on 870-meter Mt. Tsukuba. It fell far short of its target of clarifying the acidification processes that caused the three medical complaint events; however, it was Japan's first "acid rain chemistry" model. The laboratory experiments did not succeed any better.

There were two sets of medical experiments to determine the cause of the medical complaints. The first was formally conducted under the auspices of the Moist Air Pollution Investigation Committee, and the second was a rather curious independent effort whose results the committee drew upon. The formal experiment was a simple one carried out on three rabbits. Various different solutions of H^+, SO_4^{2-}, NO_3^-, NH_4^+, and HCHO simulating precipitation concentrations were tested on the rabbits. The solutions were equivalent to, or up to ten times stronger, than the concentrations found in precipitation samples collected in the city of Kumagaya in Saitama prefecture, which was the site of the greatest number of medical complaints in 1975. The solutions were applied to the rabbit's right eye and the left eye was left untouched. Observations for injurious symptoms (e.g., bloodshot eyes) were made. The results of all tests were negative. No symptoms were found in any of the rabbits at any of the concentrations.

The second experiment was controversial. In 1975, the first year of the investigation, a group at the Kanagawa Prefecture Pollution Research

Center published a one-page brief in the proceedings of the sixteenth annual meeting of the Air Pollution Society describing an experiment conducted on human patients (Kurokawa et al. 1975). Their results suggested that compounds such as aldehydes (specifically, formaldehyde and acrolein)[8] may have caused the effects. The experiment was conducted as follows. Kurokawa and her colleagues first collected samples of precipitation from three locations in Kanagawa Prefecture. (Kanagawa Prefecture is located just west of Tokyo and includes the port city of Yokohama.) They analyzed the samples for pH, formaldehyde, and acrolein.

They found high concentrations of the aldehydes, with a maximum formaldehyde value of 1.9 ppm, and a maximum acrolein value of 0.23 ppm. Based on their collected data, in the second part of their analysis, they applied to the rim around the eye solutions of various pH, formaldehyde (0–10 ppm), and acrolein (0–1 ppm) values. Their results showed that low pH solutions without aldehydes did not manifest any irritation. However, the frequency of eye irritation increased when the formaldehyde level was 3.0 ppm or above, and the acrolein level was 0.3 ppm or above, even at higher pH values. Their overall conclusion was that the cause of eye irritation in the 1974 Kantō Plain incident could not be attributed simply to the acidity in precipitation, that various aldehydes were also involved.

The experiment is simple enough. However, the puzzling aspect was that the authors did not report who the patients were, how many they applied their test solutions to, or where this experiment took place. Nor were the results published in more detail in any later paper. Nor were there any follow-up experiments. To me, this was surprising given that the results strongly suggested that acidity alone was not the culprit in the medical complaints, and that Kurokawa et al.'s results were often referenced in later works. I remained puzzled until I was told that the patients in the experiment were the researchers themselves, and that they were ordered by superiors to cease such experiments.

Moist Air Pollution Investigation Evaporates

In the end the cause of the medical complaints that initiated the moist air pollution investigation was never discovered. The Final Report concluded: "Only a fragmentary understanding of the incorporation of irritant pollutants into precipitation, especially mist/drizzle, and of the formation of such pollutants in water droplets was gained. Numerous problems must remain unsolved" (Photochemical Formation of Secondary Pollutants Investigation Committee—Moist Air Pollutants Subcommittee 1981, 249; author's translation). The baffled committee also had to admit that for unknown reasons the phenomenon that initiated the project completely disappeared of its own accord. And here the investigation terminated. (Even today it is not known exactly why the phenomenon disappeared, though the most likely explanation is that the stringent air pollution control measures undertaken in the 1970s incidentally controlled the causative agents of the moist air pollution problem.)

During its tenure the Moist Air Pollution Investigation Committee was not influenced to any significant degree by foreign acid rain research. The character of the Japanese moist air pollution problem was sufficiently different from then current European and North American acid rain problems that foreign experiences were not directly translatable to the Japanese situation. In the language of this book, there were no "universal bridging objects" that the Japanese could draw upon. They were on their own. Thus, Japanese researchers, while acknowledging the existence and potential applicability of foreign acid rain research, did not employ the same monitoring equipment, chemical models, analytic techniques, etc. The Final Report shows a clear awareness of acid deposition events in Europe and North America, though. For example, the European Monitoring and Evaluation Programme (EMEP) established in 1977 is noted, as is the pioneering work of Svante Odén, the discoverer of the problem in Europe, and the work of Gene Likens, the discoverer of the problem in North America.

The final chapter of the Final Report, titled "Future Work," portends a shift toward Western-style analysis of acid deposition, a shift that actually took place some five years later. Specifically, the committee recommended the following:

1. Precipitation chemistry monitoring methods should be changed from separate volume sampling to bulk sampling (monthly, or per rainfall event). The report notes that this was the sampling method used by

EMEP. In addition, the committee recommended that Japanese monitoring be standardized as in EMEP.

2. A nationwide and year-round monitoring network should be established. Up to this point systematic monitoring had taken place only in the Kantō area during a two-week period in the rainy season. It was also suggested that in the future an international network in Asia might be necessary. (This was the first mention of the possibility of an Asia network; a network that is today a reality.)

3. Research on ecological impacts to soils, plants, lakes, rivers, and wetlands should be conducted, as well as research on damage to materials.

4. Finally, the report recommended more research on emission sources, transport and diffusion processes, chemical conversion from primary to secondary pollutants, the incorporation of pollutants into precipitation, and dry deposition processes, and recommended development of a long-range transport model.

After Moist Air Pollution

After the moist air pollution investigation ended, piecemeal research continued at local government environmental research institutes. However, before we discuss this, some noteworthy research that took place independent of local government efforts is highlighted.

Independent Research

1978: Precipitation pH in Tokyo

Katsuko Saruhashi and Teruko Kanazawa (1978) measured the pH of rainfall at the Tokyo location of the Meteorological Research Institute of the Meteorology Agency over a five-year period from July 1973 to July 1978. They measured pH and rainfall amount for every rainfall event—a total of 364 events over the five-year period. They found the five-year average pH to be 4.52. This they compared to the 4.1 pH value for Tokyo reported by Miyake in 1939. They noted that Miyake's results showed significant seasonal variation between summer and winter, but that such variation was no longer seen by 1978. Saruhashi and Kanazawa further compared their 1978 results on dissolved inorganic components in precipitation to similar research for three different years (1939, 1948, and 1964) that spanned a forty-year period. They found that (1) the sulfate concentrations fell somewhat, (2) the nitrate concentrations rose dramatically, and (3) chlorine ion concentrations remained the same. Their paper is valuable as a comparative analysis of historical data; however, as was all too common with Japanese papers of this and earlier eras, no theoretical insights were offered and no attempt was made to tie the findings into a larger environmental picture. In addition, and again like most other papers, it was poorly referenced. There is tantalizing mention of the existence of older works but they are not included in the simple four-item bibliography.

1978: Acid Fog and Forest Damage in Hokkaidō

Unusual defoliation in pine plantations outside of the industrial city of Tomakomai on the southeastern coast of Hokkaidō was observed for the first time around 1970. Investigation into the causes was begun in the late 1970s by Yoshitake Takashi, and the first results were published in the annual report of the Hokkaidō branch of the Forestry and Forest Products Research Institute (Yoshitake 1978). Damage to *pinus strobus* was observed every year during the July to August growing season. A preliminary investigation ruled out salt damage from winds off the sea, and ruled out snail or moth infestation. However, several measurements of pH and the sulfate content of rainfall and fogwater revealed low pHs and high sulfate concentrations, thus suggesting the need for further analysis of the effect of high acidity during the growing season. A more extensive investigation was undertaken (Yoshitake and Masuda 1980). In addition to *pinus strobus*, damage to two other pine species and six broadleaf species was analyzed, and more extensive pH and sulfate measurements were made. Defoliation and damage was found in all species, mean pHs of between 4.1 and 4.3 were measured during the two-month growing season, and sulfate concentrations over 100 ppm were occasionally observed in the precipitation.

The full-blown results of the investigations conducted between 1977–1980 were not published until 1986 (Yoshitake and Masuda 1986). The area of damage extended from 3 to 20 kilometers inland from the coast around Tomakomai, and from 40 to 250 meters above sea level.

Above 250 meters, trees were healthy. Defoliation and damage to both pine and broadleaf species during the July–August growing season coincided with the season of fog formation around Tomakomai. Wind trajectories carried coastal fog from the industrial areas toward the affected region up to a height of 250 meters. Fogwater samples frequently revealed pHs of 4.0 or lower during the growing season, and high sulfate concentrations, with one reading of over 800 ppm.

The leading investigator in the preceding research, Yoshitake Takashi, now at the Forestry and Forest Products Research Institute in Tsukuba, commented in the final paper on the intensive inquiries into acid rain-related forest ecosystem impacts in Europe and North America. There were no reports of acid rain damage to forests in Japan up to this time. Yoshitake did not directly attribute the damage discovered in the Tomakomai region to acid rain/fog. This is in sharp contrast to the case of Sekiguchi Koichi (described later and in the next chapter), who, though not a forestry specialist, straight out claimed that dieback of Japanese cedar in the Kantō area was due to acid rain. Sekiguchi published his conclusions the same year, 1986, as Yoshitake's more circumspect conclusions.[9]

1981: pH 2.86

Previous to his forest-related work, Sekiguchi in 1981 recorded a rainfall event in Maebashi in Gunma Prefecture near Tokyo with a pH of 2.86 (Sekiguchi et al. 1983). The event occurred on 26 June 1981 and was estimated to have covered a total area of 36 square kilometers. The rainfall was very light, more of a mist, with a total rainfall amount of only about 1 millimeter. The sulfate concentration was 46.8 ppm, and nitrate 95.3 ppm. Over a two-week period, sixty rain samples had been collected and analyzed. The 2.86 event was one of these samples. This event attracted media attention because it was one of the lowest recorded since the moist air pollution investigation ended.

1983: Comparison of Tokyo Precipitation Composition

In 1980–1981 Okita Toshiichi and Komeiji Tetsuhito repeated analysis of the chemical composition of rainwater in Tokyo conducted by Okita

Table 7.1
Historical comparison of the chemical composition of rainfall in Tokyo (1962–1963 and 1980–1981)

	Ratio of 1962–1963 to 1980–1981 Ionic Composition of Rainfall in Tokyo									
	Ca^{2+}	Mg^{2+}	Na^+	K^+	NH_4^+	Cations	SO_4^{2-}	NO_3^-	Cl^-	Anions
Ratio	4.4	12.2	1.5	2.4	1.4	3.4	3.1	0.65	1.1	2.2

Source: Okita (1983, 550).

in 1962–1963 (Okita and Komeiji 1983). It revealed some interesting changes over the twenty-year period. A summary of their results is shown in table 7.1.

If the ratio given in the table (1962/1963 value ÷ 1980/1981 value) is greater than 1 it means that an ionic species decreased in Tokyo in the twenty-year interval, and if the ratio is less than one the ionic species increased. The table shows that: (1) the cations, particularly Ca^{2+}, Mg^{2+}, and K^+ significantly decreased, (2) SO_4^{2-} decreased by one-third, and (3) NO_3^- roughly doubled. The authors offer only a few partial explanations for the changes. They speculate that strict SO_2 controls and the declining use of coal in the 1970s reduced the sulfate levels, and that an increase in percentage of roads paved reduced the alkaline species' levels.

Local Government Research

During the five years of the moist air pollution investigation both the national and local governments were involved, but after its demise continuation of the work reverted to local governments. By this time many research institutes outside the Kantō area had joined in collecting data. In the ten-year period between 1974–1984 at least thirty prefectural institutions, sampling in at least 300 locations, analyzed precipitation chemistry in Japan. Thus, even after termination of the investigation in 1979, precipitation chemistry knowledge continued to accumulate. Precise comparison of data was difficult, though, not only because the local government research institutes were stuck on the separate volume sampling method of the moist air pollution investigation (it is hard to compare species concentrations when the volume of the sample is extremely small), but also because there was no uniformity or coordination of sampling methods among the various institutions. Despite this, the local government data became the jump-off point for the EA's new acid deposition research program begun in 1983, which is described in the next chapter.

By far the most comprehensive summary of acid deposition research by local government environmental institutes between 1974 and 1984 is contained in a special issue of the journal *Kankyō Gijutsu* (Kankyō Gijutsu 1985). This issue contains articles on local research in Hokkaidō, Niigata, Miyagi, Chiba, Gunma, Kanagawa, Nagano, Aichi, Fukui, Kyoto, and Nara prefectures, and the Tokyo Metropolitan Area. It also contains an outstanding review article by Tamaki Motonori (Tamaki, 1985) on the research at these and other locations in Japan up to 1984. Tamaki's article contains a list of more than 300 references. All, with only two inconsequential exceptions, are cites to Japanese journal articles by Japanese authors.

The following are conclusions about precipitation chemistry in Japan drawn from the ten-year research results obtained by local government environmental research institutes.

• The range of pH values was between 3.0–6.0, and the average was about 4.5. By 1984 it was still not possible to accurately determine large-scale, volume-weighted pH patterns in Japan because sampling methods and pH determination methods varied significantly.

• Over the ten-year period there was no obvious overall downward trend of pH values. Some areas showed a downward trend, such as Yokkaichi, Kitakyushu, and Kumamoto, but most showed a stable pattern. However, low pH values (<4) appeared with more frequency in the later years than the earlier years.

• It used to be that low pH values occurred during the winter, but beginning sometime in the 1970s this seasonal trend flipped and low pH values began occurring during the summer. This reversal seemed to be due to the decrease in coal use for winter heating and the increase in vehicle traffic generating NO_x emissions. However, though dominant, this trend was not uniform throughout Japan. Some areas still showed a winter low pH, and other areas show little seasonal variation.

- Generally, low pHs were associated with light rains.
- Some of the lowest recorded pH values between 1973–1981 are shown in table 7.2.

Perhaps the most important conclusion, based on the finding that precipitation chemistry measurement in remote locations were similar to metropolitan areas, was that acidification was widespread and not limited to big cities.

Analysis of Period 4

The most significant development during Japan's fourth acid deposition period was scientific and political recognition of an acid rain problem (the moist air pollution problem). An epistemic shift from the ill-defined *acidified atmosphere* problem-framework of the third acid deposition period to a *moist air pollution* problem-framework distinguishes Japan's third and fourth periods. Unlike previous periods, Western scientific ideas related to atmospheric chemistry did not propel the shift; a domestic event in Japan did—the outbreak of medical problems associated with low pH rainfalls in 1974.

In response, the Environment Agency established the Moist Air Pollution Investigation Committee. This was Japan's first formally constituted acid deposition expert community. It conducted a five-year program of research; however, the committee failed to discover the cause of the problem, and after 1979 research efforts devolved to local government environmental research institutes. Local institute research helped reestablish an imperative for policy action, though. In summary, between 1974–1983 there were two stages to the creation of acid deposition-related scientific knowledge in Japan: (1) the moist air pollution investigation (1975–1979), which focused on the health problem in the Kantō area; and (2) local environmental research institute efforts (1979–1983), which basically imitated the type of research initiated by the moist air pollution investigation but expanded it to new locations. During both stages scientists attempted to construct a moist air pollution problem-framework.

Table 7.2
pH values <3 as measured by local government research institutes from 1973–1981

pH	SO_4^{2-} [ppm = mg/l]	NO_3^- [ppm = mg/l]	Place	Date	Comments
2.8	8.3	3.5	Shizuoka City	1973, June 28	
2.7			Fujinomiya	1973, June 29	
2.9	11	1.8	Yokohama	1977, Aug 25	
2.9			Sahara	1978, May 11	
2.9	6.2	68	Oigawa	1980, Aug 7	
2.6			Kagoshima	1981, June 11	Associated with volcanic activity
2.8	25	19	Omaezaki	1981, June 20	
2.6	18	8.7	Omaezaki	1981, June 22	
2.9	47	95	Maebashi	1981, June 26	Total rainfall 1 mm; see description of this event above
2.9	3.9	10	Hamamatsu	1981 June 30	

Source: Based on table 3 in Tamaki (1985, 138).

Moist Air Pollution Problem-Framework

In 1974 Japanese scientists did not discover an acid rain problem as much as an acid rain problem discovered them. Acid-related medical injuries to thousands of people prompted a hastily assembled committee of scientists to investigate the headline-grabbing news. The problem did not fit any previously defined problem-framework, so the committee set out to create one.

As Japan's first acid rain document, written in 1975, clearly indicates, Japanese scientists were well aware of acid rain research then being conducted in Europe. However, since the problem in Japan centered around human injury, not an ecological injury, the European research was not directly relevant. Perhaps the most outstanding innovation in Japan was invention of new monitoring techniques. In Europe, bulk samplers were employed to gather long-term data on total precipitation (samples were commonly collected over a one-month period), whereas in Japan a separate volume sampler was employed to gather short-term, fractional volumes (collecting separately the first millimeters of precipitation). The two monitoring methods yielded different types of information—bulk sampling gave data on total acid inputs to ecosystems, and separate volume sampling gave data on incremental, small-volume acid fractions that might irritate surfaces on the human body during very fine mists and drizzles. In general, Japanese scientists combined new and old methodologies related to precipitation chemistry, computer modeling, and impacts research. Armed with these methodologies, they gathered data and then set about trying to develop a theoretical understanding of the problem. In other words, they attempted to construct a compelling problem-framework. They failed.

It is not surprising that the Moist Air Pollution Investigation Committee failed in its mission, for instead of encountering a neat and orderly environmental problem it met with a very messy one. It ran smack into the full, ugly mix of air pollutants enveloping the Tokyo urban-industrial megalopolis. This situation stood in sharp contrast to Europe and North America. First in Europe, then shortly after in North America, a relatively neat and orderly problem was discovered—namely, long-distance transport of sulfur compounds that acidified rain that in turn acidified freshwaters and soils when deposited on them.

The object of attention in Europe and North America was basically a single chemical end-product—sulfate. In the Kantō area, by contrast, sulfate and nitrate played roughly equal roles (with nitrate probably dominant). In addition, aldehydes, other organic acids, ammonia, and ozone vastly complicated the picture. All of these pollutants were essentially bundled together in the problem tackled by the Moist Air Pollution Investigation Committee. In Europe it wasn't until the 1990s that these extraneous chemicals entered the picture. Witness, for example, the progression of protocols to the Long-Range Transboundary Air Pollution Convention—sulfur (1985), nitrogen oxides (1988), volatile organic compounds (1991), and persistent organic compounds (1998).

In Europe the acid rain problem was first discovered in "remote" Scandinavia, which imported long-lived sulfur compounds (versus the much shorter-lived nitrogen compounds). In the Scandinavian case, at least as initially understood, the other pollutants such as encountered in the Kantō area were conveniently eliminated in long distance transport, leaving only sulfur compounds and thus creating a relatively simple problem to analyze (relative to the Kantō problem).

In short, the Moist Air Pollution Investigation Committee confronted a problem that was far too complex for the state of research at the time. Thus, not only was the problem not understood in scientific terms, but no policy debate nor policy emerged. Why? Political incentive to deal with the problem died first and foremost with the mysterious disappearance of the problem, and secondly with the inability of the committee to derive convincing science-to-policy bridging objects. Their problem-framework lacked a cause-and-effect chain of reasoning, and it lacked an "impact message." By default, this limited scientists' abilities to create conceptual science-to-policy bridging objects. The one science-to-policy bridging object to emerge was "low pHs in the Kantō area," but this wasn't enough to sustain political interest. A potentially powerful cause-and-effect bridging object was "rainy season low pH and/or high aldehyde concentrations causes bodily irritations," but there was insufficient scientific evidence to back it up. The moist air pollution investigation did, however, succeed in setting up a large-scale, organized precipitation chemistry monitoring network (which in itself is a material versus conceptual science-to-policy bridging object) and in further developing scientists' technical skills in acid deposition science. This experience nourished the historical scientific momentum on the acid deposition problem.

The committee's partially formulated moist air pollution problem-framework, in spite of its deficiencies, did yield a set of science-to-science bridging objects (e.g., separate volume sampling methodology) that resulted in its dissemination to local research institutes. After 1979 these science-to-science bridging objects, despite the absence of a centrally coordinated research program, perpetuated the framework pioneered by the committee. During this second phase to moist air pollution research, a theoretical framework continued to elude scientists. The fractured and fragmentary problem-framework that emerged between 1979 and 1983 remained a patchwork. However, a significant science-to-policy bridging object did emerge—"low pHs around the country." The problem-framework was not sufficiently extensive to make a nationwide assessment of the state of the acid rain problem but it did have sufficient geographical coverage to be able to state that low pHs were being recorded in multiple locations around the country.

During the fourth acid deposition period, Japan's first coherent community of acid deposition scientists emerged, initially formed around the moist air pollution investigation. The community was bonded by a common interest in the problem of acidification of the atmosphere, and consisted of a total of some twenty to thirty researchers by 1983, almost all of whom devoted only part-time to this type of research. The leading figure was Okita Toshiichi. Others included Hara Hiroshi of the National Institute for Public Health, Tamaki Motonori of the Hyogo Prefectural Institute of Environmental Science, and Komeiji Tetsuhito of the Tokyo Metropolitan Research Institute for Environmental Protection. The expert community was the vehicle for spread of the moist air pollution problem-framework, primarily through publications, scientific meetings, and direct personal contact. Its active members were also eventually responsible for persuading the government to again pursue acid deposition research. This story will be picked up in the next chapter.

Moist Air Pollution Policy

The one major incidence of policymaking during the fourth acid deposition period was the initial decision to set up and fund the Moist Air

Pollution Investigation Committee. This is an example of what might be called "targeted science policy," where a government makes a conscious decision to support scientific research on a specific topic. The decision to establish and fund the committee was made within the Environment Agency; impetus derived essentially from the crisis atmosphere that still pervaded Japan after the domestic environmental revolution.[10] Public clamor essentially demanded attention to any sudden and new environmental problem. The Moist Air Pollution Investigation Committee's study was just one of dozens of investigations into the causes of various environmental problems that cropped up around this time. As already stated though, no regulatory policy nor policy debate emerged from the moist air pollution investigation. The minimal science-policy interaction during the fourth period is illustrated in figure 7.1. Even though the moist air pollution investigation was not a generator of political debate, its successor, the Acid Deposition Survey (to be explained in the next chapter), is a tremendous generator of political debate.

Acid Rain in Europe and North America

Before leaving the fourth acid deposition period, we will pause to briefly review the science and politics of acid rain as it was unfolding in Europe and North America, for they were the source regions for the ideas that inspired Japan's next acid deposition period.[11] The U.N. Conference on the Human Environment, or Stockholm Conference, held in 1972 placed the acid deposition issue on the international political agenda in Europe. Also in 1972 Norway established the SNSF (Sur Nedbors Virkning på Skog of Fisk, or Acid Precipitation Effects on Forests and Fish) Program, a massive eight-year, U.S.$16 million endeavor that was, and still is to date, the largest research program ever undertaken in Norway.[12] Through this and other research, Norway and Sweden added substance to the European problem-framework set forth by Svante Odén. Also in 1972 the Organization for Economic Cooperation and Development (OECD) launched a four-year review (1972–1976) of the problem of long-range transport of air pollutants. At the end of the review (OECD 1977), the OECD concluded that: (1) long distance transport was indeed occurring, (2) the area of acidic deposition comprised the majority of northwest Europe, and (3) of the eleven countries participating in the

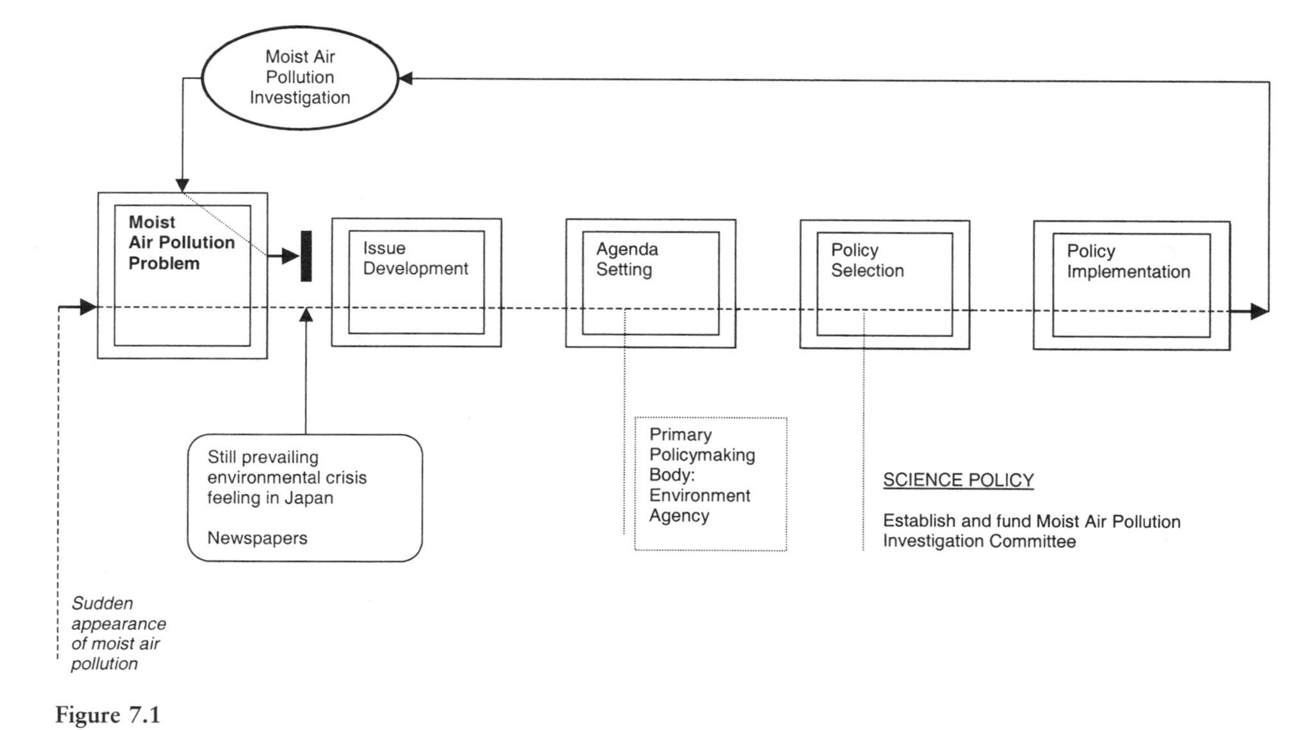

Figure 7.1
Science-policy interface of the moist air pollution problem (1974–1983)

study, five (Austria, Finland, Norway, Sweden, and Switzerland) received more transboundary pollutants than domestic, and six (Belgium, Denmark, France, West Germany, The Netherlands, and the United Kingdom) received more domestic than transboundary pollution. Despite significant uncertainty in the data and analysis, the OECD study nevertheless provided the first independent scientific evidence to back up Swedish and Norwegian claims of the existence of transboundary transport of air pollution. As a result of the OECD study, the European Monitoring and Evaluation Programme (EMEP) was established in 1977. Thus, by the mid- to late-1970s a compelling international acid rain problem-framework existed in Europe, from which, as we have seen, Japanese scientists drew upon to map the future direction of Japanese acid deposition research after the failure of the moist air pollution investigation.

In Europe the culmination of the research programs, conferences, and debates of the 1970s was the signing of the United Nations Economic Commission for Europe (UNECE)'s Convention on Long-Range Transboundary Air Pollution (LRTAP) in 1979.[13] LRTAP, however, did not initially have teeth; it was a weak treaty. Sweden and Norway were the prime movers to strengthen it, and West Germany and the United Kingdom were the prime resisters. These latter two nations essentially held veto power over concrete actions proposed by the Scandinavians. But in the early 1980s, discovery of *Waldsterben* (forest death), first confirmed in West Germany, changed the entire picture for Germany, as well as for the rest of Europe.[14] The discovery of widespread damage to forests in Europe placed forest impacts on par with aquatic impacts in the European problem-framework. At a 1982 conference the West German government announced a dramatic turnabout in its attitude toward international controls on transboundary air pollution, and from this point on joined the Scandinavian nations as leaders in promoting stringent air pollution/acid deposition international policy measures. One result was that in 1984, only one year after LRTAP came into force, the first protocol to the Convention (on financing for EMEP) was adopted.

Scientific and policy movement in North America was slower than in Europe. The first scientific effort to map out the scale of the acid deposition problem in North America was the Canadian Network for Sampling Precipitation (CANSAP), established in 1976. This was followed by the U.S. National Atmospheric Deposition Program (NADP) also established in 1976. In 1980 Canada and the United States signed a Memorandum of Intent to conclude an acid rain treaty (which didn't actually happen until 1991). Also in 1980 the U.S. National Acid Precipitation Assessment Program (NAPAP), the largest acid deposition research effort in the world, was established. Despite the auspicious start in North America, after the election of Ronald Reagan in 1980, obstruction by the new administration in the NAPAP process seriously affected the science and strained otherwise friendly relations between the United States and Canada over transboundary air pollution. A frustrated Canada became increasingly furious in the early and mid-1980s over U.S. inaction on the issue. Scientific investigation of acid deposition continued, however, in both the United States and Canada and basically reinforced the European problem-framework that emphasized impacts to sensitive ecosystems.

Sustainability

During Japan's fourth acid deposition period, the swirl of events in the West were peripheral to Japan's activities. Japan, however, first recognized acid rain as a distinct problem during this period, and recognized it as an element in its overall sustainability debate. But because the manifestation of the problem (human health versus ecological damage) was so different than in the West, Japan at first took an entirely different approach. To structure its approach, Japan relied on its domestic scientific capability and knowledge base. Scientists succeeded in hewing a rough problem-framework; however, no solution path emerged and the human health aspect of Japan's acid rain problem subsided as a sustainability issue. This set the stage for wholesale importation of the ecologically oriented Western problem-framework. As we shall see in the next chapter, this brought ecological sustainability into Japan's thinking on acidification of the atmosphere for the first time. In the West, acid rain was already labeled by many "one of the most serious environmental threats of our time."[15]

8

Period 5 (1983–1990): Ecological Acid Deposition Research

Acid Rain Surveys

1983: Environment Agency's Acid Rain Survey

Because of the scientific and political attention the acid rain problem drew in the West, and because local government research continually reported low pH rainfalls throughout Japan, the EA decided to reinstitute an investigation of the acid deposition problem.[1] Kato Saburo, then director of the Air Quality Bureau of the EA, described the process by which the Survey came into being as follows.[2] After completing work on adding NO_x to the total mass emissions control system through an amendment to the Air Pollution Control Law, there was breathing space to pursue other projects. Researchers such Okita Toshiichi and Hara Hiroshi of the National Institute for Public Health, Tamaki Motonori of the Hyogo Prefectural Institute of Environmental Science, and Komeiji Tetsuhito of the Tokyo Metropolitan Research Institute for Environmental Protection had been recommending that a national level acid rain research effort be established.

Kato took up the idea and a plan for a survey was drawn up within the Air Quality Bureau. The plan was circulated among higher officials in the agency and approval obtained. The survey was included in the EA's budget request for fiscal year 1983. The next step was consultation with the Liberal Democratic Party (LDP), for the ruling party had to approve a ministry's budget before it could proceed to the next stage. The LDP approved the EA budget, one item of which was the Acid Rain Survey. This cleared the way to take the budget to the Ministry of Finance.

In the meantime, a prominent article appeared in the *Asahi* newspaper on the acid rain problems in the West and in Japan. Armed with this article, which Kato relates carried a great deal of weight because it demonstrated public concern about the problem, an EA representative met with the official within the Ministry of Finance who was in charge of overseeing the EA's budget. This official essentially had line-item veto power over anything in the budget request. The Ministry of Finance official, however, was apparently impressed by the importance of the issue and approved the survey, funding it at a level of about U.S.$100,000 per year.

In 1983 the Acid Rain Survey (*Sanseiu Taisaku Chōsa*, literally, Acid Rain Countermeasures Survey) was established. A Committee on Acid Rain (*Sanseiu Taisaku Kentōkai*, literally, Acid Rain Countermeasures Investigation Committee), headed by Okita Toshiichi, was the governing body for the Survey. Designed as a modest effort, it was, however, the first nationwide survey to employ identical sampling procedures and the first to study ecological impacts. To date three phases are complete: phase 1 (1983–1988), phase 2 (1988–1993), and phase 3 (1993–1998). The methods and results of the Survey-phase 1 are discussed later. While the Survey was being conducted, other research related to acid deposition also flourished. This research is discussed first.

1984: CRIEPI's Inland Sea Regional Acid Deposition Project

In 1984 the Central Research Institute of the Electric Power Industry (CRIEPI), the research arm of the electric power industry, began a one-year Inland Sea Regional Acid Deposition Project headed by Fujita Shinichi. Although it was preceded by earlier research in the late 1970s (Fujita and Terada 1985), the Inland Sea Regional Acid Deposition Project was the first to specifically address the acid deposition problem. Between June 1984 and May 1985 data was collected from a twenty-station monitoring network laid out in a grid pattern on the island of Shikoku in the Inland Sea (*Setonaikai* in Japanese) to determine regional-scale rainfall chemistry and wet deposition. Results were published in CRIEPI (1985) and Fujita and Kawaratani (1988).

Mid-1980s: Acid Rain and Forests

During the 1980s a flurry of research activity on forest decline and its potential relation to acid deposition appeared. It was stimulated by the surfacing of numerous reports of damaged trees (generally isolated stands, not forests as a whole) around Japan, especially in and near urban areas, and by the ominous warnings emanating from Europe, Germany in particular, on extensive forest damage seemingly attributable to acid deposition. In Japan widespread tree decline was reported in the following areas:

1. Kantō Plain around Tokyo
2. Mt. Akagi in Gunma Prefecture
3. Oyama in the Tanzawa Range in Kanagawa Prefecture
4. Tomakomai in Hokkaidō
5. Kansai (Kobe-Osaka-Kyoto) and Inland Sea area

Only research in the Kantō Plain area surrounding Tokyo is discussed here because it was by far the most influential both scientifically and politically. Research on Tomakomai was discussed in the previous chapter.

Sekiguchi et al. (1986) published a report concluding that damage to *sugi* (Japanese cedar or *cryptomeria japonica*) over a large area, more than 5,000 square kilometers, west and northwest of Tokyo in the Kantō Plain was due to high acid deposition and/or photochemical oxidants. This was the first public announcement attributing tree damage ("grand scale dieback"—in the authors' words) to the influence of acid deposition. *Sugi* is the most abundant tree species in Japan. It grows almost the entire length of the Japanese archipelago, and occupies a special place in Japanese culture (as discussed in chapter 3). This in part explains why the Sekiguchi paper created a considerable stir.

The authors conducted a one-year survey of damage beginning in the spring of 1984. Over fifty sites in two prefectures (Gunma Prefecture, a damaged area, and Ibaraki Prefecture, a nondamaged area) were selected for observation and at each site the number of Japanese cedars, their height, diameter, and state of damage were recorded. More than 80% of the trees of 15 meters in height and 40 centimeters in diameter were heavily damaged. The team then looked at five potential causes of the dieback: (1) temperature, (2) precipitation, (3) nature of soil, (4) acid deposition levels, and (5) photochemical oxidant concentrations, and compared each factor in damaged and nondamaged areas.

They first concluded that temperature and precipitation were not significantly different between the two prefectures over the past decades, and that the soil types were basically the same for both prefectures. Then examining the pollutant levels they found the following, based on data from fifteen national air pollution monitoring stations in the two prefectures: (1) the annual average concentration of SO_2, and NO_x were about two times greater in Gunma, than Ibaraki, prefecture, (2) the H^+, SO_4^{2-}, and NO_3^- wet deposition rates were also about two times greater in Gunma than Ibaraki, and (3) the number of days photochemical oxidants exceeded 0.06 ppm was five to ten times more in Gunma than Ibaraki prefecture. These data were consistent with the fact that the prevailing winds blow pollutants from the Tokyo metropolitan area toward Gunma Prefecture, especially in summer. Based on these facts the authors concluded that of the five factors "acid deposition and oxidant [sic] are the most probable causes of decline" (Sekiguchi et al. 1986, 268).

Sekiguchi et al. could not get their paper accepted by a Japanese journal so it was published in Britain. Because of the stir it caused, a series of more detailed surveys in the Kantō area appeared. Yasushi Morikawa of the Forestry and Forest Products Research Institute conducted a survey in 1985–1986 (Morikawa 1989; Morikawa et al. 1990). The survey clarified the area of damage as confined to the central, north/northwest area of the Kantō Plain at elevations less than 200 meters. This, Morikawa and his fellow researchers observed, was strikingly different from the general patterns found in Europe and North America where it was high elevation stands that seemed to be faring worst. They found that the area of damage correlated closely with distributions of SO_2, NO_2, and O_3, but that the distribution of no one pollutant could explain the distribution of damage. They pointed out that, during the latter part of the 1960s, rainfall in the central Kantō Plain was much less than normal, and that this dry period may be one factor contributing to the decline. In addition, there was no indication of increasing acidity or aluminum concentrations (aluminum can be mobilized by acid inputs causing injury to root systems) in the soil. Therefore, they concluded that there was no evidence that acid deposition was a dominant cause of forest decline in the Kantō Plain.

Takahashi Keiji and his colleagues (Takahashi et al. 1986) also conducted a survey of the Kantō Plain beginning in 1984. In general, their results agreed with those of the Forestry and Forest Products Research

Institute. They found that the areas of decline in the Kantō Plain coincided well with the distribution of high oxidant levels and low amounts of precipitation in the growing season. Takahashi et al. (1991) followed up their Kantō study with another in the Inland Sea area in 1986–1987. Again, they found that high oxidant and low precipitation levels during the growing season correlated with the areas of tree decline. This added weight to their hypothesis that these were the major factors in the decline.

In none of these cases discussed here could forest decline unequivocally be attributed to acid deposition. Certainly air pollutants seemed to be involved, but whether it was primary pollutants (such as sulfur dioxide) or secondary pollutants (such as sulfuric or nitric acids, or photochemical oxidants) could not be ascertained.

1987: Effect of Acid Rain on Agricultural Crops

Kohno Yoshishisa of the Terrestrial Biology Department of CRIEPI completed a series of research experiments on the effects of acid rain on agricultural crops by 1987. The first experiments, on the effect of acid rain on the growth and yield of radish and kidney bean plants, were published as CRIEPI reports (Kohno and Fujiwara 1981, 1982). These experiments showed that leaves exposed to acid rain with a pH of 3.0 or below showed necrotic spots, and that the plants experienced reduced growth.

Following these initial experiments, a more extensive set of experiments ensued in 1985 and extended over a three-year period to 1987. In these experiments soybean seedlings were exposed to simulated acid rain using an automatic rain-generating system in a greenhouse. The results were first reported in a CRIEPI report (Kohno 1987), and later appeared as two articles in the English journal *Water, Air, and Soil Pollution* (Kohno and Kobayashi 1989a, 1989b). Based on their results the authors concluded that current ambient levels of rain acidity in Japan with a mean pH of about 4.6 "should not have an adverse impact on seedling growth in soybean" (Kohno and Kobayashi 1989a, 11), and "would not directly affect seed production of . . . soybean" (Kohno and Kobayashi 1989b, 173). These results were consistent with those published by CRIEPI's counterpart in the United States, the Electric Power Research Institute (EPRI). EPRI concluded that then-current levels of rainfall acidity in the United States did not adversely affect growth or yield of agricultural crops.

1987: Meteorological Society of Japan's Special Report on Acid Rain
The Meteorological Society of Japan (1987) published a special report on acid rain that provides a glimpse of the state of acid deposition knowledge in Japan at the time. The 170-page report contains sections on the history and general state of acid rain knowledge in Japan and the West. A total of ten authors contributed. The report demonstrates several interesting points. First, it highlights the fading influence and relevance of the health impact-oriented moist air pollution research of the late 1970s and early 1980s. This type of research is mentioned only in passing. Second, it illustrates the surprising degree of absorption of knowledge from the West, for the special report is primarily a summary of Western knowledge. One indication is that of the total number of citations in all the bibliographies of the special report, 212 are to Western papers and only 67 are to Japanese papers. And third, the report exposes the weak state of integration of research underway in Japan. All research was basically stand-alone. It is not integrated between research projects. Besides the Meteorological Society of Japan report, other compendia of Western acid deposition-related knowledge and techniques appeared in the 1980s (see, for example, the discussion on air pollution models next).

Mid to Late 1980s: Atmospheric Modeling
Research on regional-scale atmospheric transport computer models related to acid deposition began in the West in the 1970s, and by the 1980s several highly sophisticated models emerged, among them RADM (Chang et al. 1987), STEM (Carmichael et al. 1986, 1991), ADOM (Venkatram et al. 1988), and RAINS (Alcamo et al. 1990). The first computer modeling of meso-scale pollutant transport (100–200 kilometers) in Japan appeared in the mid-1980s.[3] The first application of a sophisticated computer model to meso-scale pollutant transport in Japan was conducted by a U.S.-Japanese team using the STEM model (Chang et al. 1989a, 1989b, 1990).

Meanwhile, Japanese scientists also evaluated dozens of models that had been developed in Europe and North America. The first reviews of individual European or North American modeling programs appeared in the Japanese literature in the early 1980s. However, the largest such effort began in 1986. The Industrial Pollution Control Association of Japan, IPCAJ (*Sangyō Kōgai Bōshi Kyōkai*), now the Industrial Environmental Management Association of Japan, IEMAJ (*Sangyō Kankyō Kanri Kyōkai*), launched a massive two-year literature search in which more than 1,000 references were gathered. Over sixty European and North American models (short-range, long-range, short-term, long-term, photochemical oxidant, and acid deposition) developed between 1973 and 1985 were compared on the basis of essential modeling characteristics (Okamoto and Katatani 1988). This work had considerable influence in introducing Western regional-scale atmospheric transport computer modeling to Japan. CRIEPI (in 1987), IEMAJ (in 1988), and Okita Toshiichi (in 1987) began developing acid deposition models.

1990: CRIEPI's Long-Range Transport Monitoring

In 1986 CRIEPI instituted another short-term monitoring project in northwestern Kyūshū, again headed by Fujita Shinichi (Fujita 1990). The project came on the heels of the Inland Sea study discussed previously. CRIEPI monitored precipitation chemistry at three sites in northwestern Kyūshū for a one-year period between June 1986 and May 1987. A comparison of wet deposition during the cold season (six months from October through March) to that of the hot season (six months from April through September) showed that wet deposition of SO_4^{2-} during the cold season constituted 50% of the annual total, despite the fact that it represented only about 30% of the total annual precipitation.

Furthermore, total emissions and deposition were estimated within a 100-kilometer radius encompassing the three monitoring sites. The emission values, in 10^9 grams sulfur per half-year, were anthropogenic (7–19) and seasalt emissions (1), and deposition values, in the same units, were wet deposition (20), dry deposition of SO_2 (10–20), and dry deposition of SO_4^{2-} (1). Although the contribution from other natural sources (biologic and volcanic) were unknown, the above estimates—$8–20 \times 10^9$ grams sulfur emissions per half-year versus $31–41 \times 10^9$ grams sulfur deposition per half-year—seemed to indicate deposition far exceeded emissions. Since much of this excess occurred during the winter, and since northwesterly winds blowing from the Asian mainland prevail during the winter, Fujita speculated that the high cold season depositions originated from emission sources on mainland Asia.

1989/1990: Acid Rain Survey—Phase 1 Final Report
The EA's Committee on Acid Rain published its final report on the results of phase 1 of the Acid Rain Survey in Japanese (EA 1989) and English (EA 1990). Phase 1 was divided into three categories of study: precipitation chemistry analysis, effects on rivers and lakes, and effects on soils. The methodology and results in each category are discussed briefly.

Precipitation Chemistry The first and foremost accomplishment of the Survey—phase 1 was establishment of the National Acid Deposition Monitoring Network (NADMN)—a nationwide, long-term, bulk sampling monitoring network. Up to this time the numerous precipitation chemistry studies conducted by various organizations used different sampling periods and different sampling devices. NADMN began operation in September 1983 at fourteen sites (one urban and one rural site in each of seven prefectures) throughout Japan. In 1985, fifteen more sites were added. Many of the sites were located at preexisting stations of the National Air Pollution Monitoring Network (NAPMN). They were equipped with bulk samplers that remained open during both dry and wet periods. Continuing the work begun by the Moist Air Pollution Committee, there was also established in the Kantō area a dense network of twenty-four sites. Thus, by the end of the Survey—phase 1 there were twenty-nine nationwide and twenty-four Kantō area sites. All sites used the same type of equipment, either filtered bulk precipitation collectors or snow collectors. The results obtained from the nationwide monitoring network are outlined next.[4]

The annual volume-weighted, station-specific mean pH ranged from 4.4 to 5.2, with a national average mean of 4.7. The final report concluded that the "spatial distribution of pH did not show significant difference[s] over the country, although some sites in . . . western Japan tended to have lower pH values than those [in] northern Japan" (EA 1990, 2–3). No significant pH trends were detected over the survey period. A number of noteworthy findings concerning ionic concentrations and total (dry + wet) deposition rates were observed.

• Nonseasalt sulfate (nss-SO_4^{2-})[5] concentrations were higher on the Sea of Japan coast than on the Pacific coast. The largest wet deposition rates were found on the Sea of Japan coast and on the island of Yakushima located at the very southern tip of Japan. Yakushima recorded the highest deposition rate—7.4 gm/m^2/yr in 1987. In both areas deposition peaked during the winter. This was attributed to local emission sources on Yakushima, but local sources could not account for the high winter values on the Sea of Japan coast. The report concluded that "the influence of emissions from local sources as well as those from the Asian Continent should be evaluated" (EA 1990, 8).

• The equivalent ratio of NO_3^- to nss-SO_4^{2-} (N/S) was maximum in Tokyo (0.83), and, excluding Tokyo, ranged from 0.19 to 0.57, with a mean of 0.30. Assuming that the precipitation acidity originates from either sulfuric or nitric acid, it points to the conclusion that the acidity of precipitation in Japan was primarily from sulfate.

• NO_3^- concentrations were highest in the Tokyo area (roughly double or more that of other sites). Also the highest rate of nitrate deposition was recorded in Tokyo—3.5 gm/m^2/yr in 1986.

The phase 1 final report compared precipitation chemistry data from Japan to that obtained by the National Acid Precipitation Assessment Program (NAPAP) in the United States. Although a detailed comparison was not possible because NAPAP used wet-only samplers whereas Japan used bulk (wet + dry) samplers, some interesting and suggestive points were highlighted. In northeastern North America many areas experienced pHs on the order of 4.0; however, in Japan the lowest annual mean pHs were around 4.4 (Tokyo area). Nss-SO_4^{2-} deposition rates in Japan generally ranged from 2 to 5 gm/m^2/yr, which was roughly comparable to that of eastern North America. High NO_3^- deposition rates of 2 to 3 gm/m^2/yr in some urban areas of Japan were comparable to those in eastern North America.[6]

Effects on Rivers and Lakes In order to determine if acid deposition was affecting the water quality of lakes in Japan, 133 lakes in fifteen prefectures were examined. During the period from 1983–1988, no obvious acidification due to atmospheric acid deposition was found in the 133 lakes. The frequency distribution of lake pH values was bimodal with one tall peak around pH 7.0–7.5, and a short peak around 4.5–5.5. There were fifteen lakes with a pH less than 5.5, all of which were thought to have been acidified by their proximity to either volcanoes or mines.

Effects on Soils Two types of tests for actual and potential adverse effects of acid deposition on Japanese soils were conducted: (1) monitoring of soil pH and cation exchange capacity, and (2) simulated acid rain experiments of plant growth and microorganism activity. Each is explained briefly.

Soils at twelve nonagricultural sites in four prefectures located the length of Japan were analyzed for soil pH and exchangeable cations over the five-year period. The pH of the surface layer was measured in situ and ranged from 4.3–6.8, with about 70% falling in the range of 5.0–6.0. No observable acidifying trend was observed over the five years. Soil samples were also analyzed in the laboratory to determine cation exchange rates. Based on the various tests and a knowledge of inherent soil characteristics, soils were tentatively classified according to their susceptibility to acid deposition. This work formed the foundation for constructing a preliminary "Map For Assessing the Susceptibility of Japanese Soils to Acid Precipitation."

In plant growth experiments, buckwheat was grown in soils from each of the twelve sites and subjected to either simulated acid rain or a control rain of distilled water. Germination was unaffected in all soils; however, growth (plant height) was slightly depressed by a pH 4 rain, and significantly depressed by a pH 3 rain. Calcium in the buckwheat decreased with increasing acidity and there was an increase in exchangeable aluminum in the soil.

Analysis of Period 5

An epistemic shift from the moist air pollution problem-framework of the fourth acid deposition period to a true acid deposition problem-framework marks the transition between Japan's fourth to fifth periods. "Low pHs around the country"—the primary science-to-policy bridging object to emerge from the moist air pollution problem-framework—was instrumental in causing the shift. During the fifth period, Japan imported and localized the rapidly developing "standard" acid deposition problem-framework of the West. Similar to the fourth period, the major policy decisions were research-related decisions. The most important of these policies was to establish, then continue and expand, the Acid Rain Survey. Between the time of the formation of the Committee on Acid Rain in 1983 and the completion of the first phase of the survey in 1988, a coherent, localized acid deposition problem-framework emerged in Japan. The final report of phase 1 in 1989 was not only the first official, nationwide assessment of the problem, but also the first document spelling out the framework.

A solid expert community of about seventy-five scientists contributed to this assessment and to other acid deposition-related research. They came from national and local government research institutes, universities, and CRIEPI. Most members were policy inactive (i.e., they concentrated on their research), but a few were policy active in that they strongly pushed the government to set up and maintain nationwide research programs. The most influential community members were those serving on committees within the Acid Rain Survey and those heading CRIEPI's acid deposition program. Their primary mode of influence was channeling information, and an emerging set of bridging objects, to bureaucrats in the EA and MITI via reports and meetings, and to the public at large via the mass media.

Acid Deposition Problem-Framework

Establishment of the Acid Rain Survey facilitated full absorption in the 1980s and 1990s of the Western acid deposition problem-framework (its methodology, content, and orientation to research). The survey not only served as the primarily vehicle for transfer of the universal elements of the Western framework but also built on Japan's indigenous experiences with acid deposition (i.e., its "historical scientific momentum"). The survey was not, of course, the only avenue of knowledge transfer and absorption. In addition, work at local research institutes, CRIEPI, and universities expedited the transfer and construction of Japan's problem-framework. Western ideas (i.e., science-to-science bridging objects) were imported through scientific journals and reports, attendance at conferences, visits abroad by Japanese scientists, and visits to Japan by foreign researchers. Once imported, Japanese scientists applied the ideas locally. The Meteorological Society of Japan's 1987 Special Report on Acid Rain and IEMAJ's 1988 study of Western computer models, both discussed previously, provide a window on this localization process.

Once localized, a set of conceptual science-to-policy bridging objects emerged. Perhaps the most important was an extent-and-intensity bridging object—"nationwide low pHs" (and its close cousin, "nationwide high sulfate and nitrate deposition values"). It was based on the results of the survey's nationwide, long-term, uniform precipitation chemistry monitoring network. During the five-year period (1983–1988) of the first phase, annual mean pHs of 5 or below were recorded throughout the country, with slightly lower values in the south and west. Annual deposition rates of nss-SO_4^{2-} and NO_3^- were at the same or higher levels than those in Europe and North America. Another important (trend) bridging object was "no overall downward trend of acidification." No significant changes in acidification of the atmosphere were detected during the five years of the survey. Other monitoring efforts, such as those of local research institutes and CRIEPI, basically supported the results of the Survey.

In addition, a series of impact or risk assessment bridging objects also emerged, three carrying low-risk messages and one a high-risk message. The three "low-risk" bridging objects spoke to lake acidification, soil acidification, and agricultural crop damage. Lake quality seemed unaffected by acid deposition. A survey of 133 lakes in fifteen prefectures revealed most lakes to have a pH of about 7. There were a number of lakes with pH of less than 5.5, but these seem to have been acidified by volcanic activity and mine drainage. Soil quality also seemed unaffected by acid deposition. No highly acidified soils nor any downward trends in the pH of soils were observed. Other research such as that at CRIEPI corroborated this conclusion. Simulated acid rain experiments by the Acid Rain Survey and CRIEPI showed that the effects on agricultural crops were insignificant. In one review of such research the author stated that "significant effects on agricultural crops and tree seedlings at current ambient levels of acid precipitation (pH 4.0–5.0) are unlikely . . . [but] . . . acid precipitation may be one of many environmental factors contributing to forest decline" (Noguchi 1990, 312). The "high-risk" bridging object related to forest damage. Impacts to forests was the most controversial area of Japan's fledgling problem-framework. Numerous areas reported forest decline. Debate on these declines caused scientific sparks to fly. Some claimed acid deposition (alone or in combination with photochemical oxidants) caused forest damage; others claimed there was no solid proof.

For the first time in Japan's acid deposition history we see knowledge related to acidification phenomena becoming contested. This is most evident in the controversy over the causes of forest decline. Thus, we begin to see the problem-framework differentiate into a relatively uncontested core (e.g., the existence of nationwide low pHs) and a contested periphery (e.g., the potential impacts of nationwide low pHs on, for instance, forests). While tremendous progress was made between 1983 and 1990, a reliable and integrated picture of acidification problems still eluded Japan. In other words, the quality of the framework still suffered. By 1990 the work had only begun. However, the existence of a sufficiently coherent problem-framework resulted in creation (by scientists) of a set of bridging objects that could span the science and policy worlds. We now turn to their political influence.

Acid Deposition Policy

Government decision-making revolved around research, not regulatory policy, during the fifth acid deposition period. At the beginning of the period in 1983, scientists, armed with the bridging object "low pH rainfalls around the country," could argue by analogy to the Western acid deposition problem-framework that Japan's ecosystems might be affected by the low pHs. However, this bridging object alone is insufficient to explain government decisions. As we saw in the case of the Acid Rain Survey, a confluence of events went into the decision to create the survey—free time within the Air Pollution Control Section of the EA, the raucous debate surrounding acid rain in the West, and an eye-catching newspaper article, in addition to the nationwide low pH bridging object.

No regulatory policy was forthcoming, however, during the fifth period. One reason was the inability of scientists to actually establish a strong, Japan-specific cause-and-effect bridging object linking low pHs to ecosystem impacts. Even though precipitation throughout most of Japan was acidic, no adverse ecological effects to inland waters or soils were detected, nor was there conclusive evidence linking acid deposition as the cause of forest decline. Some soils and inland waters were considered potentially at-risk, and this fueled further research, but not regulatory measures. Also, that forests were being damaged was not in question, but that acid deposition was the prime suspect was indeed in question. This again spurred research but not regulatory action.

Figure 8.1 illustrates the science-policy interface during the fifth period. In the figure, starting with the lower, left-hand ellipse, the policy process more or less chronologically followed the arrows until reaching the "no regulatory policy" block. The dominant actors in policymaking were scientists and bureaucrats. There was strong mutual feedback between the two. Science-to-policy bridging objects during this period were often initially expressed in the executive summaries of scientific reports prepared for bureaucrats, and then reinforced during face-to-face meetings between elite, policy active scientists and lower level bureaucrats. From this point of origin they would disseminate to other social worlds.

As the bridging objects emanated out of the scientific world, other actors interpreted their content, often against the background of certain nature-related elements of Japanese culture (which strengthened the political implications for some and weakened them for others). Industry was not a major actor during the fifth acid deposition period. Even though industries that emitted acidic pollutants were put on the defensive by the acid deposition issue, they used the lack of definitive impacts to argue against further crackdown. Also, they could tout their air pollution control measures implemented starting in the 1970s as acid deposition mitigation measures.

Manufacturers of control technology relished the opportunity to further promote their products. Hitachi, for instance, came out with a wonderful full-page advertisement showing a picture of a *mikan* (Japanese orange), lemon, and *umeboshi* (Japanese pickled plum), each of which has a pH of around 4.5, 2.5, and 2.0, respectively.[7] The ad says

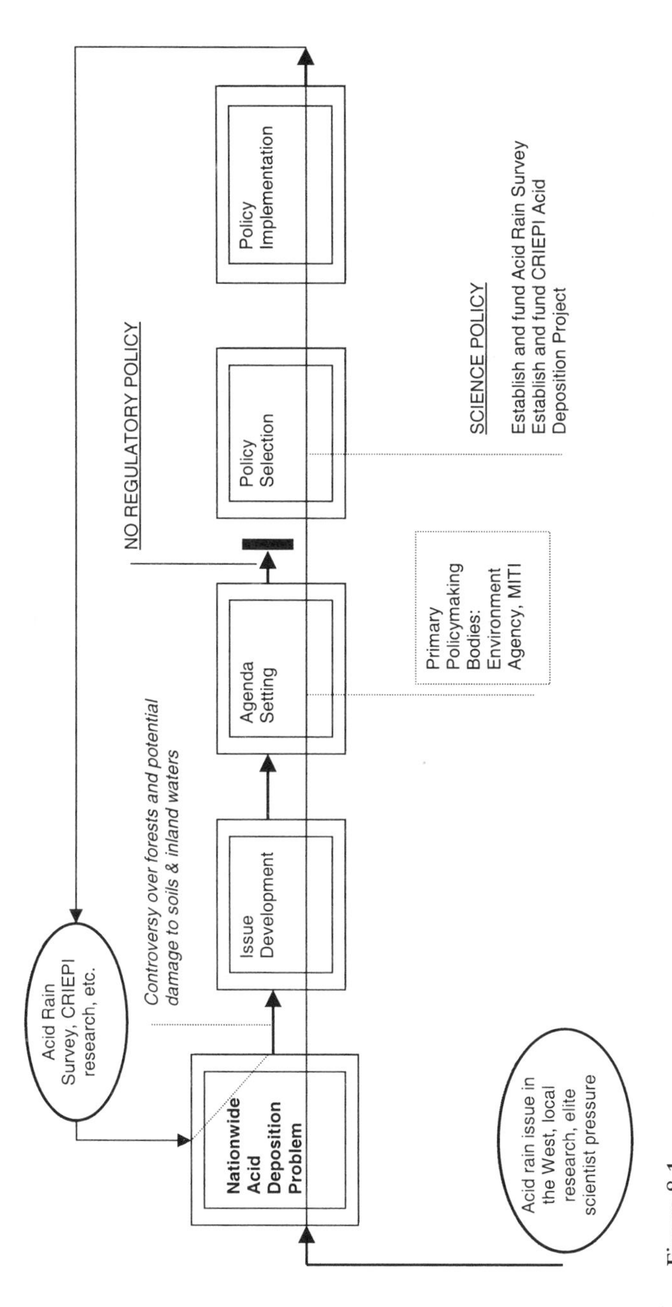

Figure 8.1
Science-policy interface of the acid deposition problem (1983–1990)

the Earth is heading toward the pH of an *umeboshi* but that Hitachi's automotive NO_x reduction technology will help prevent this from happening. Other actors, such as LDP and Diet members, while involved in aspects of decision making, were not central players in the development of research policy.

The general public, in loose alliance with the mass media, was an interesting influence on policy. A high level of awareness about the acid deposition issue developed among the Japanese public during the 1980s. One indication is a citizen acid rain monitoring boom that began in the late 1980s (Tamaki 1993). Tens of thousands of school children and citizens took part in acid rain monitoring events.[8] These events received widespread media coverage. Samples were collected in everything from kitchenware to complicated sequential collectors, and analyzed with everything from simple color indicator kits to sophisticated hand-held pH meters. The public's deep interest in the issue can in part be attributed to Japan's "culture of rain" and "culture of wood."[9] I was told repeatedly in discussions with ordinary citizens that rain and forests are special to the Japanese people. Thus, for instance, the "high potential risk of forest damage due to acidification" bridging object picked up heightened urgency when interpreted by many political actors, especially the general public. In contrast, because Japan is not a "culture of lakes" the "low potential risk of lake acidification" bridging object picked up little or no added urgency when seen through a cultural lens. Ultimately, the public's acid deposition-related activity and awareness, while not directly shaping policy, provided a constant reminder to the central government to be vigilant on the problem.

By 1990, just as Japan was settling the parameters of its domestic acid deposition debate, evidence was appearing that set the stage for the explosive conversion of the problem from a domestic to an international one. This story is told in the next chapter. Meanwhile, Europe and North America had forged ahead scientifically and politically on their respective international problems.

Acid Deposition in Europe and North America

While Japan was absorbing in the 1980s the standard problem-framework developed in the West, events in Europe and North America were leading to the development of important new science-to-policy bridging objects. In Europe during the 1980s acid deposition science and policy evolved rapidly and became tightly linked. The link, though, was tenuous at the beginning of the decade. There was yet considerable antagonism between scientific research findings and policymakers willingness to enact control measures. West Germany and the United Kingdom opposed stringent controls. However, the scientific revelation of *Waldsterben*, or forest death (or more properly, widespread forest decline in Europe) helped induce both countries to reverse their position. Adoption of the first pollutant-specific protocol to LRTAP (the Helsinki Sulfur Oxide Protocol) occurred in 1985.

A nitrogen oxides protocol was signed in 1988. And a framework for integrating the latest scientific research with decision making was established under LRTAP through EMEP and LRTAP's Working Groups, Task Forces, International Cooperative Programs (ICPs), and Meetings of Acidification Research Coordinators (MARCs).[10] Today there are Working Groups and Task Forces on technology, economics, the atmosphere, effects, development of critical loads, and integrated assessment modeling. There are ICPs on forests, fresh waters, materials, and crops. There is a Coordinating Center for Effects based in The Netherlands, and two coordinating centers for EMEP in Oslo and Moscow. Not only did a highly integrated picture of acidification problems emerge during the 1980s in Europe, but with it a highly integrated program of scientific research feeding into a relatively unified process of international policymaking. The net effect was development of a dense network of science-to-policy bridging objects, many of which eventually influenced Japan.

A master bridging object developed in Europe during the 1980s was the concept of "critical loads."[11] As adopted by the Executive Body of LRTAP in 1988 critical loads are "a quantitative estimate of an exposure to one or more pollutants below which significantly harmful effects on specified sensitive elements of the environment do not occur according to present knowledge" (Swedish NGO Secretariat on Acid Rain 1995, 2). Essentially this defines an environmentally acceptable load of a pollutant to which an ecosystem can be exposed without harmful effects.

The critical loads concept was not invented in Europe. Its first appearance was in Canada in 1985 (Brydges and Wilson 1991, 5–6). However, the Scandinavians latched onto the idea and turned it into a functional scientific-political concept. The Nordic Council began promoting the idea of environmentally defined deposition values as the basis for emission reductions (instead of flat-rate reductions) in 1986. Up to this point, emission reductions were essentially political agreements relying on social and economic pressure for acceptance rather than on the capacity of ecosystems to handle acidic deposition. The critical loads approach forces negotiators' attention on scientific issues. In 1988 a Working Group was appointed to study the critical loads approach, and in 1991 the Executive Body approved of the approach as an acceptable, effects-base, scientific method for devising strategies for air pollution abatement.[12]

Critical Loads are the central element of the policy-oriented, integrated assessment computer model known as the RAINS (Regional Acidification, Information, and Simulation) model (Alcamo et al. 1990; Hordijk 1995). The model itself is a second master bridging object developed in Europe. It was developed at the International Institute for Applied Systems Analysis (IIASA) in Austria beginning in the early 1980s. The intent was to create a model that was codesigned by experts and users, that was of modular construction, that was simple yet based on more detailed models or data, and that was user-friendly. The resulting PC-based model uses energy and emissions data from EMEP, atmospheric transport and deposition data produced by EMEP's atmospheric dispersion models, the critical loads maps developed by the Coordination Center for Effects, and a cost model, including an extensive abatement technology database, developed at IIASA. The RAINS model is now used to perform calculations for all acidifying pollutants. The first protocol to use the model, and thus incorporate the critical loads concept, was the Oslo Sulfur Dioxide Protocol adopted in 1994. The critical loads concept and the RAINS model, more than any other bridging objects, have brought science and policymaking into intimate proximity in Europe. Some call this a revolution in transboundary air pollution management.

In North America the situation was not quite so pretty. Little in the way of innovative science-to-policy bridging objects were developed. The decade of the 1980s started on a high note (signing of the 1980 MOI and authorization of NAPAP), suffered through a dismal low (Reagan administration meddling in the NAPAP scientific process), and ended on a high (signing of the 1991 Canada–U.S. Air Quality Accord). The scientific showpiece in North America was unquestionably NAPAP (1990). However, political interference with scientific results sullied the top notch science performed under this program and inhibited construction of science-to-policy bridging objects. The environmental picture revealed by NAPAP was one of a few obviously negatively impacted areas (e.g., acidified lakes in the Adirondacks, and damaged high altitude red spruce in the northern Appalachians) amidst a sea of potentially stressed ecosystems. But this impact message did not capture policymakers' nor the public's attention.

Canada, on the other hand, while playing an activist role in North America like Scandinavia in Europe, was more or less forced to follow the scientific and political momentum generated by the United States. Canada's equivalent to NAPAP, the Canadian National Acid Rain Research Program, did not get off the ground until 1985 and did not compare in scope or complexity to NAPAP. Canada published its final assessment report on acid deposition in the same year as publication of NAPAP's Integrated Assessment (RMCC 1990). Throughout the decade, the Reagan administration opposed any concrete measures to mitigate transboundary transport of pollutants. The political climate improved with the election of George Bush in 1990. Passage of the U.S. Clean Air Act soon after his election resolved many of Canada's complaints about U.S. inaction.

The most significant event related to environmental management of acid deposition in North America was the development of an emissions trading program in SO_2. While highly innovative, emissions trading is *not* a science-to-policy bridging object. It would fall under the category of an economics-to-policy bridging object. With the passage of the 1990 Clean Air Act, the United States embarked on the world's largest use of a market-based instrument to attain environmental protection. To date, the SO_2 emissions trading program is highly successful. From an ecological point of view, its main defect is that the national emissions ceiling that caps the number of permits traded is not based on ecological criteria; it is determined by political negotiation with only indirect input from science. The emissions trading (economic) bridging object invented in North America has had little influence on Japan's and East Asia's acid deposition policymaking as compared to the major influence of the European critical loads and RAINS model (scientific) bridging objects.

Sustainability

Compared to Europe and North America, Japan was just getting out of the starting blocks on acid deposition research during the 1980s. Scientists, however, succeeded in constructing a problem-framework and a cogent set of bridging objects. Lay-translated outlines of the framework soon began appearing, such as in the EA's "Quality of the Environment in Japan" annual reports. And introductions by scientists could be found in bookstores; see, for instance, Taniyama (1989) and Murano (1993). In addition, acid rain became part of the curriculum in secondary schools and a citizen monitoring movement began.

All of this activity laid the foundation for entry of acid deposition into Japan's contemporary sustainability debate. The key development for addressing acid deposition sustainability in Japan during the fifth acid deposition period was establishment of a nationwide, ecologically oriented scientific assessment process designed to ascertain acid deposition sustainability. Out of this assessment process emerged a broad brush solution path—reduction of domestic emissions of acidic substances would increase the pH of rainfall and reduce the likelihood of negative impacts to sensitive ecosystems, especially forests. But before controversy over this solution path got off the ground, the whole terms of the debate changed.

9

Period 6 (1990–present): East Asian Transboundary Air Pollution

Japan's fifth environmental era (1990–present)—the global environment era—is characterized by the emergence of international and global issues on Japan's environmental agenda. It coincides with Japan's sixth, and present, acid deposition historical period, one characterized by the internationalization of its acid deposition problem. During this period, acidic pollutants were harbingers leading to general recognition of transboundary air pollution in East Asia. While cross-boundary transport of a wide variety of air pollutants is now acknowledged in the region, acidic pollutants were, and still are, the center of scientific and political attention. The international dimension of Japan's acid deposition problem (i.e., import of externally generated acidic pollutants) eclipsed the domestic dimension (i.e., impact of internally generated pollutants). This chapter begins with a description of Japan's present-day environmental era, the larger context within which the sixth acid deposition period is situated. A discussion and analysis of the sixth acid deposition period follows through three phases: (1) discovery of the transboundary air pollution problem (mid-1980s–1990), (2) initiation of international cooperation (1990–1995), and (3) establishment of EANET, the East Asian Acid Deposition Monitoring Network (1995–2001). The latter two phases express Japan's foreign policy emphasis on regime formation. The chapter closes with a discussion of progress toward forming an East Asia transboundary air pollution regime and its relation to acid deposition sustainability.

Fifth Environmental Era (1990–present): Global Environment

During its fourth (1973–1990) environmental era, Japan suffered from a poor international environmental reputation.[1] It was labeled an "eco-outlaw,"[2] a "whale-killing, forest-stripping bogeyman on the environmental stage,"[3] and a country with a "pioneer-society mentality,"[4] which exploited resources with no thought to the consequences. The West saw Japan as a nation insensitive to international environmental problems whose government acted as a noncooperative, noncomplying member of international environmental treaties.

Some of the most widely publicized transgressions included whaling (from the mid-1950s to the present Japan has been the world's largest whaling nation), tropical hardwood logging (Japan was, and still is, the world's largest importer of tropical hardwood), trade in endangered species (Japan was the world's largest per capita consumer of wildlife and wildlife products and the largest trader in endangered species up to the 1990s), drift-net fishing (in the 1980s Japan was the largest user of the largest drift nets for deep sea fishing), foreign aid for development (Japan was heavily criticized for certain environmentally destructive foreign aid projects), and export of pollution (Japan was accused of exporting polluting industries to less developed countries where the same strict pollution standards found in Japan were not followed).

There were, of course, areas where Japan's reputation was positive, such as its cooperation on the 1987 Montreal Protocol on stratospheric ozone and its strong stance on CO_2 reductions and support for a global climate treaty, but these did not outweigh the negative. Japan in the 1980s labored through heavy criticism of its international environmental policies. This took a positive turn, however, around 1990. The change was slow in coming but is of such significance that I have chosen to define Japan's present environmental era in terms of its newfound global environmental awareness. Around 1990 a number of events took place that heralded the new era, including:

- *1988: Discovery of long-range transport of air pollutants in Asia* In the late 1980s the fact of long-range transport of air pollutants from continental Asia to Japan was confirmed. All of a sudden Japan, which previously considered itself relatively immune from the pollution problems of its neighbors, found itself in the position of "pollution victim." (The rest of this chapter discusses this discovery and its implications.)

• *1989: Council of Ministers for Global Environmental Conservation* This council of Cabinet members was established to improve coordination among ministries and agencies on matters related to the global environment.

• *1990: Action Program to Arrest Global Warming* A twenty-year program to reduce domestic and foreign greenhouse gas emissions was established.

• *1990: G-7 Houston Summit* At the Houston Summit of the G-7, Japan announced its "New Earth 21—Action Program for the Twenty-First Century." This is a MITI-controlled, big-budget, long-term project with a none-too-modest goal to "undo the damage done to the earth over the past two centuries since the industrial revolution." It is technologically oriented, aiming to develop a new generation of environmental technologies.

• *1991: Keidanren's "Global Environment Charter"* Keidanren (Federation of Economic Organizations), the country's most powerful economic lobby, published its "Global Environmental Charter." It contained environmental guidelines for Japanese corporations operating abroad.

• *1991: Eco-Asia* Japan formed Eco-Asia, an annual forum of environmental ministers from nations in Asia and the Pacific. The first meeting was held in 1991.

• *1992: Host to CITES* The eighth Conference of the Parties to the Convention on International Trade in Endangered Species (CITES) was held in Kyoto. This was the first time a major international nature conservation conference took place in Japan.

• *1992: Northeast Asian Conference on Environmental Cooperation* Japan helped organize the first of an ongoing series of conferences of officials and experts from China, South Korea, Mongolia, and Russia who gathered to discuss environmental issues in Northeast Asia.

The preceding list is by no means complete, but it indicates that Japan by the early 1990s had set its sights on transforming its laggard status to a leadership position in global environmental problem-solving. During this whirlwind of international activities, domestic environmental issues persisted, but they were now overlaid with keen attention to global environmental issues.

In 1993 the OECD completed its second environmental performance review of Japan (OECD 1994). (The first review was conducted in 1976.) The report praised Japan for decoupling economic growth from increasing SO_2, NO_x, and CO_2 emissions, and energy use. This was achieved by very strict standards, use of best available technologies, a technology-forcing regulatory approach, and high public and private investment in pollution control. Overall, OECD rated air pollution control high; waste management practices good; water quality poor; nature conservation improving; urban amenities still deficient; and increasing private consumption patterns and dependence on the automobile worrisome. The report also praised Japan's efforts in international environmental cooperation. It concluded that Japan "is playing a strong and expanding role in solving international environmental problems" (OECD 1994, 194). It also noted that Japan was one of the first countries to create a government structure to deal with global environmental issues (the abovementioned Council); that it was the world's largest ODA donor; and that it was actively promoting scientific exchanges, technical training, and technology transfers. The report did not mention the acid deposition problem but did state in the very last sentence that "the potential damage from transboundary pollution and global pollution from outside its borders, the impact of its activities on the natural resources and ecosystems of other countries, and its economic capacity and technological know-how are compelling reasons for Japan to be deeply engaged in international environmental affairs" (OECD 1994, 195).

A variety of reasons help explain the emergence of Japan's global environmental awareness and leadership. First, Japan was responding to a general trend throughout the world of recognizing global environmental issues. Second, Japan's economic superpower status by the late 1980s led to demands that it assume environmental responsibilities commensurate with its economic power. Third, as a result of the high standards of living achieved ordinary citizens began to acknowledge values other than economic success. Fourth, Japan saw opportunities to export its environmental technology. And finally, the discovery of long-range transport of air pollutants from the Asian mainland to Japan triggered a reassessment of its ecovulnerability. It is to the scientific discovery of transboundary transport of acidic pollutants and its political implications that we now turn.

Internationalization of Acid Deposition

Around 1990 East Asia (Japan, China, North and South Korea, Taiwan, far eastern Russia, and Mongolia) emerged as the world's third regional-scale transboundary air pollution hot spot behind northern Europe and eastern North America.[5] A new, internationally oriented acid deposition problem-framework that incorporated and superseded the previous domestically focused problem-framework appeared in Japan. In other words, there was an epistemic shift to what is referred to here as a long-range transport, or LRT, problem-framework. The appendix provides an extended introduction to the methodologies, knowledge content, and orientation of this framework. A distinguishing feature compared to all other historic frameworks is the diminishing influence of Western acid deposition science. By 2000 Japan's acid deposition science was roughly on par with that of Europe and North America. Japan is still the only country in Asia that can make this claim.

During the 1990s there was a sharp distinction between policymaking on the domestic and international dimensions of the problem. This reflected the nature of the acid deposition problem as it was perceived in policymaking circles. Generally speaking, domestic acid deposition was not considered a stand-alone problem, whereas acid deposition from emission sources outside of Japan was considered a stand-alone problem. The domestic problem was folded into debate on a menagerie of acidic and nonacidic air pollutants. As of 2000, high ambient air pollutant concentrations in urban areas, including smog precursors and airborne toxics from waste incineration, especially dioxin, dominated domestic debate. However, policymaking on transboundary transport of air pollutants focused primarily on acidic pollutants, especially SO_2 and SO_4. Also, whereas the domestic acid deposition problem was not a driving force in Japanese domestic environmental politics, the international problem was (and still is) a driving force in international environmental politics, especially regarding international scientific and technological cooperation. At the domestic level, there was no perceived urgency to formulate policy explicitly targeting the domestic acid deposition problem in and of itself. The reason, as stated in the last chapter, is essentially because there was no solid evidence of large-scale ecological or nonecological damage in Japan due specifically to domestic acidic emissions. The impetus for formulating domestic regulatory policy related to acidic substances arose indirectly from their connection to continuing ambient air quality problems, such as high NO_x and O_3 concentrations in urban areas, not acid deposition impacts per se. If large-scale acid deposition impacts due to domestic emissions become a confirmed reality in the future, then the legislative, administrative, and regulatory structure already set up for combating air pollution problems will almost certainly be the forum for further controls on acid deposition precursor species.

Because the international dimension of acid deposition superceded the domestic dimension during the 1990s, and because the most innovative decisions were made relative to the international dimension, I focus almost exclusively in this chapter on Japan's foreign policymaking. Japan's goal became formation of a transboundary air pollution regime in East Asia (i.e., a signed international agreement to control pollutant emissions in the region). By the early 1990s, the two main prongs to Japan's regime formation quest were clear—(1) establishing an East Asian monitoring network, and (2) extending technical aid to China to reduce its emissions. The key questions we seek to answer in this chapter are: What foreign policy decisions were made to promote regime formation? Who made them and why? And what was the role of science in influencing these decisions?

Starting with the "who" question, acid deposition-related foreign policymaking, as with domestic policymaking, follows a model that has been called "bureaucratically patterned pluralism."[6] This model postulates that a variety of interest groups influence policymaking; however, their voices are directed through channels specified by the bureaucracies. Historically in Japan, as in China and Korea, the bureaucracy has long been a preeminent element in the political system.[7] Continuing this tradition, in postwar Japan the bureaucracy has held a virtual monopoly over the legislative process (hence scholars developed "bureaucratic dominance" models of decision making in Japan). Neither the National Diet (legislative branch of government) nor the courts (judicial branch of government) had much power over the bureaucracy. The various bureaucracies were not only the drafters but also the principal interpreters and implementers of law. Very often in the formulation of a bill the major fights occurred over which bureaucracy was given jurisdiction over what territory. Once the territory was defined the bureaucracy had great freedom to act within that sphere. Within a bureaucracy the decision-making process was, more often than not, less than transparent, in part because most decisions were made through an ill-defined process of consensus. The superior position of bureaucracies was reinforced by the fact that bureaucrats were an elite class, and almost all were educated at the most prestigious universities, particularly Tokyo University.

There is no question that central government bureaucracies have dominated acid deposition foreign policymaking. The Environment Agency (EA) and the Ministry of International Trade and Industry (MITI) played the leading roles. Lesser roles were played by the Ministry of Agriculture, Forestry, and Fisheries (MAFF), the Meteorological Agency (MA), and the Ministry of Foreign Affairs (MoFA).

Unexpectedly, the EA was the innovative leader in foreign policymaking and more or less "won" the battle of the bureaucracies on the international acid deposition issue during the 1990s.[8] The EA's leadership role was unexpected because it is considered a "weak" and "junior" ministry. Despite its ostensible role as the principal administrative guardian of the environment, the EA is low on the bureaucratic totem pole. The political circumstances of its creation ensured this status because the traditional ministries fought usurpation of their powers by an upstart agency, and sought to limit its enforcement powers. Thus, the agency's principal function came to be coordination rather than implementation of environmental polices. Furthermore, the EA consistently receives small budget allocations. Often the entire EA budget is a fraction of the funds allocated to other agencies for environmental activities.

Although the EA is now some thirty years old, and much has changed within the Japanese political system, the position of the EA relative to the traditional, more powerful ministries is slow to change. In 2001 the EA was elevated to the rank of ministry—the Ministry of Environment

(MoE). How this affects its standing relative to other ministries remains to be seen. The organs within the EA that are most relevant to acid deposition policymaking are the Air Quality Bureau, the Global Environment Division, and two research institutes—the National Institute for Environmental Studies (NIES), the chief scientific research arm of the EA, and the Institute for Global Environmental Strategies (IGES), the chief institute for policy analysis. The transboundary acid deposition issue was one of the EA's first forays into international environmental diplomacy.

In part the EA's preeminence in acid deposition foreign policymaking stemmed from its historical scientific momentum on the issue. The EA had been involved in acid deposition research since the Moist Air Pollution Investigation in 1974. MITI was a relative latecomer to acid deposition research. In addition, the EA had been running the nation's air pollution monitoring network since 1971 and precipitation chemistry network since 1983. The fact that establishing a monitoring network was a logical starting point for international cooperation put the EA in a commanding position. It was a stretch for MITI to claim monitoring under its purview.[9] We will return to the tension between EA and MITI again, but first we need to examine the role of other policy actors.

The power of bureaucracies in Japanese politics has slowly eroded, especially after the bursting of the so-called "bubble economy" around 1990. Nonbureaucratic actors therefore have gained influence. However, in acid deposition foreign policymaking, with the exception of scientists, nonbureaucratic actors remain in the background.

There are two segments of industry interested in the transboundary air pollution issue—companies that emit acidic pollutants, and those that manufacture control technology. Polluting companies, while generally resistant to further domestic controls, are relieved by the transboundary issue because it more or less gets them off the hook; it draws attention away from their role in the domestic acid deposition problem. Thus, these companies are generally supportive of Japan's foreign policy initiatives. Companies selling control technologies are also supportive because the government's promotion of emission reductions in Asia provides them with greater business opportunities.

Citizen groups, the general public, and the mass media exert indirect influence on acid deposition foreign policy. The mass media avidly follows the issue. Articles on China's environmental problems and on the transboundary air pollution problem appeared regularly in the 1990s. One article even referred to air pollutants from China as "environmental aerial bombs" (*kankyō kūbaku*) (*Asahi Shimbun*, 25 February 1993, 5). Few NGOs have targeted the transboundary air pollution issue, in part because Japanese NGOs are far weaker than their western counterparts. They generally do not have the resources or expertise to jump into the political fray, even if the government would listen to their positions.

One noteworthy NGO effort, however, was the three-year "Seikyō Nationwide Acid Rain Survey" conducted by the Japanese Consumers' Cooperative Union, or *Nihon Seikatsu Kyōdō Kumiai Rengōkai*, or *Seikyō* for short.[10] The survey, started in 1993, was and still is to date the largest citizen acid rain monitoring effort in Japan. It consisted of some 100 sites uniformly distributed across the country. Results not only showed transboundary pollution from Asia, but also demonstrated high deposition in the Inland Sea region where domestic emissions dominate. Seikyō's survey garnered widespread media attention.

Even though direct channels for citizen input into policymaking are essentially nonexistent, public opinion expressed via the mass media (the major vehicle for citizen input) nonetheless provides a constant reminder to the central government that its citizens, who often see the acid deposition problem through a cultural lens, want clean rain and green forests. This, in turn, bolsters bureaucrats' resolve to tackle the problem. If bureaucrats waiver, there are tens of thousands who will notice. Bureaucrats often related to me in interviews that they felt citizens were watching their every move on the acid deposition issue.

Other political actors, such as National Diet and political party members and local government officials, occasionally weigh in on the transboundary air pollution issue, but again their involvement has not been direct. They have, however, generally supported the bureaucracies' foreign policy initiatives.

The leading nonbureaucratic actors up to the present are scientists. None of the other actors wield as much influence as scientists. Why? The simple answer is that virtually all decisions on transboundary air pollution so far relate to the less contentious domain of research policy and technical assistance to Asian countries, not the more volatile domain of regulatory policy. This gives technical experts the edge. During the 1990s, it was bureaucrats working in tandem with scientists within bureaucratically determined channels that basically set acid deposition foreign policy.

My argument goes as follows. Bureaucrats have made the key decisions but within the context of a cognitive structure developed by scientists (i.e., the LRT problem-framework). Active scientists communicated this framework to bureaucrats via science-to-policy bridging objects (to be discussed). These bridging objects, set within certain political and cultural contexts, have been sufficient to define a policymaking object for bureaucrats and propel policy action with the added impetus that other policy actors (industry, citizens, local officials, etc.) have interpreted the same bridging objects in a more or less parallel manner. I am not making the argument that scientists will remain the dominant nonbureaucratic actor; they almost certainly will not. However, to date, given that most acid deposition-related foreign policy decisions have revolved around issues related to scientific research and technical assistance, scientists have wielded inordinate influence as compared to other nonbureaucratic actors. Thus, in the rest of this chapter I focus primarily on role of scientists and scientific knowledge in shaping Japan's active stance in addressing the East Asian transboundary acid deposition problem. We begin with an examination of Japan's acid deposition expert community.

Japan's Acid Deposition Expert Community

By 2000 the expert community consisted of an estimated 250 members, with an inner core of some 50 researchers, and an outer core of some 200.[11] Generally, the inner core devotes full-time to acid deposition research, and the outer core part-time. Prominent members of the inner core sit on project-related committees and on government advisory committees. Roughly half of all community members are associated with local environmental research institutes (two or three individuals per prefecture in forty-seven prefectures), fifty with national research institutes, fifty with universities, and twenty-five with other organizations such as CRIEPI. In the mid-1980s the numbers were less than half of these figures. An indication of research output is gained from the literature list maintained by Tamaki Motonori of the Hyogō Prefecture Environmental Research Institute, which as of 1997 exceeded 4,300 references (Tamaki 1997). Tamaki started his list in 1988, together with collecting a copy of each reference. Most of his collection was destroyed in the Kobe earthquake of 1995. He has since painstakingly rebuilt it. His list and collection is today the definitive compilation of Japanese-produced acid deposition scientific literature.

There is a high degree of awareness and conscious identification in Japan as a member of an acid deposition knowledge-generating community. This awareness first emerged in the late 1970s during the Moist Air Pollution Investigation. By 1980 a small handful of researchers (a dozen at most) came to identify themselves as "acid rain researchers." What bound this tiny group was the belief that acid rain was an issue of concern in Japan. They, after the demise of the Moist Air Pollution Investigation in 1979, lobbied for establishment of a nationwide, ecologically oriented research program. The resulting Acid Rain Survey (now Acid Deposition Survey) significantly increased the number of scientists engaged in acid deposition research, and caused a quantum leap in conscious identification as a member of a well-defined research community.

With the realization around 1990 that acid deposition was not only a domestic problem in Japan but also an international problem in East Asia, the community underwent another quantum leap in numbers and awareness. Then in the mid-1990s the growing community underwent yet another revolution in self-awareness precipitated when Japan was selected to host the sixth International Conference on Acid Deposition in 2000.[12] Up to this time Japan's acid deposition expert community was introverted. There was little identification as part of an international community of researchers. At the forth International Conference on Acid Deposition in Scotland in 1990 only one Japanese researcher attended. However, at the fifth conference in Sweden in 1995 over fifty researchers attended. At the sixth conference held in Japan, hundreds of Japanese researchers attended. Also, until the mid-1990s only a small fraction of acid deposition research conducted in Japan was published in English.

From interviews with researchers it is clear that up until this time they did not feel much compulsion to disseminate Japanese research results to the wider world. The community was content to produce knowledge for domestic consumption only. However, after the fifth conference a process of internationalization took hold and completely transformed the community's self-image.

Despite this internationalization, the community is not homogenous. It is divided into bureaucratically determined subgroups. A vertical structuring exists in which a scientist's lines of communication are strongest vertically within his or her home bureaucracy and weaker horizontally across bureaucracies. There is a strong aversion among bureaucracies to be dependent on another's scientific information. The two dominant programs—the EA's Acid Deposition Program and CRIEPI's Acidic Deposition Project—are parallel, large-scale research programs that hardly cross paths. There is, of course, discourse between individual scientists and information is presented by both groups at the same meetings and published in the same journals, but at an institutional level it is almost as if the other program didn't exist. There is virtually no formal cooperation between the two programs.

The vertical structuring of acid deposition science implies that the most intense scientist-bureaucrat interaction takes place between bureaucrats within a given ministry and scientists who are funded by that ministry. In particular, most day-to-day interaction takes place between middle and lower level bureaucrats in a ministry and upper level scientists at that ministry's national research institute(s). These parties basically determine the nuts and bolts of a ministry's acid deposition research agenda, e.g., research themes, personnel allocations, and project funding levels. Since the bulk of funding is funneled through national research institutes, the top scientists at these institutes wield tremendous influence in science policy. Specifically, the National Institute for Environmental Studies (NIES) of the EA, the Forestry and Forest Products Research Institute (FFPRI) of MAFF, the National Institute for Resources and Environment (NIRE) of MITI, and CRIEPI (which technically is a private [not a national government] research institute but which is included here because of its intimate ties to MITI) are the dominant scientific research institutes. They are the pacesetters of the scientific research agenda in the acid deposition field in Japan, and they are most influential in policymaking.

Does the bureaucratic segmentation of scientific research and the expert community seriously distort acid deposition foreign policymaking? In my estimation, no, at least not yet. Each of the EA and MITI's 'bureaucratic' problem-frameworks and associated bridging objects are essentially the same. Thus, to date, rather than undermining each other's position, they basically reinforce it. This has been true from the time of the discovery of transboundary air pollution, the topic to which we now turn.

Phase 1: Discovery of Transboundary Air Pollution (mid-1980s–1990)

Japanese scientists first discovered the transboundary air pollution problem in the mid-1980s. The discovery was a gentle and collective process. Coming up with the idea that pollutants could travel from mainland Asia to Japan required no great feat of imagination—the transport of dust from the deserts of Asia to Japan (the *kōsa*, or yellow sands, phenomenon) had been known for centuries, and the fact of long-range transport of pollutants in Europe and North America provided ready example of the potential in East Asia.[13] However, to actually detect the pollutants required sophisticated monitoring equipment and siting of the equipment at locations free from local pollution. No single scientist's name in Japan is associated with the discovery of long-range transport of air pollutants, as we associate the name of Svante Odén with the discovery in Europe or Gene Likens in North America. A great many Japanese scientists contributed. The EA's Acid Deposition Survey reports carried the first official accounts of their work.

Phase 1 of the Acid Deposition Survey ended in 1988. During its course the first scientific evidence for long-range transport of air pollutants in East Asia was uncovered. It came from monitoring data that showed high ambient air concentrations of sulfate aerosols and high deposition values of nonseasalt sulfate during the winter months at monitoring stations located along the Sea of Japan coast. Since northwest winds from the continent prevail during the winter, this constituted strong circumstantial proof that Japan was receiving air pollutants from outside its borders. The very first reports of this data appeared as early as 1986.[14] The first "official" mention of such data appeared in the interim report of the survey in 1987 (both a Japanese version (EA 1987a) and an English version (EA 1987b) were published). "Official" is placed in quotes because the reports do not place any special emphasis on the Sea of Japan data. They are mentioned only briefly in the Japanese version, and were edited out of the English version. The survey's phase 1 final report, released in 1989, again makes only brief mention of the possibility of long-range transport. It notes the high sulfate values on the Sea of Japan coast, but comments only that "meteorological conditions and emissions sources, including those on the continent, need further investigation" (EA 1989, 2; author's translation). Thus, in complete contrast to the explosive appearance of the international acid rain issue in Europe and North America in 1968 and 1972, respectively, its appearance in East Asia was stunningly subdued.

Further evidence for long-range transport emerged from CRIEPI's one-year monitoring project in northwestern Kyūshū in 1986–1987, as mentioned in the previous chapter. This and other early reports of monitoring evidence prompted the first efforts at developing long-range transport (LRT) computer models. The first such models were developed by CRIEPI (begun in 1987), the Industrial Environmental Management Association of Japan, or IEMAJ (begun in 1988), and Okita Toshiichi (begun in 1987). All of these models readily verified that long-range transport of significant quantities of air pollutants from the continent, especially during winter, could indeed occur, but none was capable of reliable estimates of actual quantities transported. (See the appendix for further discussion of Japan's LRT computer models.)

Unlike Europe and North America, there were no dramatic announcements, no shocking newspaper articles about a chemical war, no conspicuous press conferences. Japanese scientists were, of course, concerned about the implications of their discovery, but they focused more on gathering additional accurate and reliable data than prematurely causing a fuss. Scientific information about the transboundary problem was therefore slow in coming into public view. This is evident from Tsuruta Haruo's (1989) article in the prominent Japanese journal *Kagaku* (Science), titled "Higashi Ajia no Sanseiu (Acid Rain in East Asia)," which lists only two references related to the transboundary aspect of the problem (the rest refer to the domestic situations in Japan and China). At no time did an "activist" group within Japan's growing acid deposition expert community materialize to beat the drum about the transboundary problem and transform it into a major political issue. An "active" group, however, quietly pressed the importance of the problem to bureaucrats via newly created science-to-policy bridging objects.

The first was an "existence" bridging object; it stated the existence of a problem—"sulfate is being imported to Japan from mainland Asia." The bridging object's simple message was based on high-quality monitoring data backed by early computer simulations. This "imported sulfate" bridging object was carried to bureaucrats in the EA by way of scientists in the Acid Deposition Survey's Committee on Acid Rain, and to bureaucrats in MITI by way of the lead researchers in CRIEPI's Acid Deposition Project. It provided a 'policy object' for bureaucrats to begin deliberating. The end result was the rapid rise of transboundary air pollution (acid deposition) on Japan's foreign policymaking agenda.

Facilitating the rapid rise was the recent creation of the Council of Ministers for Global Environmental Conservation (in 1989). The Council's mission was to guide Japanese international environmental policy and improve coordination among ministries and agencies on international environmental issues. It was chaired by the Prime Minister and included the heads of nineteen ministries and agencies. The Council's existence ensured that if an issue such as acid deposition made it into its chambers, it would have the ear of the top level of government. The first Council meeting was held in June 1989. At the second meeting agreement was reached to draw up a comprehensive research program related to the global environment, and at the third meeting in June 1990 the Global Environmental Protection Research Plan (GEPRP) was approved (GEPRP is explained in more detail in the appendix). The importance accorded the acid deposition issue at this time is demonstrated by the fact that it was included as one of a handful of key "global environmental problems" receiving concentrated research attention in GEPRP.

How and why did the acid deposition issue make its rapid rise on the international research agenda? Before looking at the answer to this question, we must understand that the process by which scientific information related to long-range transport of air pollutants, in the guise of LRT bridging objects, reaches high-level decision makers has been relatively consistent during our final acid deposition period. The various bridging objects, after honing by scientists, often in the process of producing project reports, are carried by senior scientists to lower-level bureaucrats and provide a "weakly structured common identity" across the scientific and bureaucratic worlds. "Carried" generally means pronounced in executive summaries of project reports, presented at project meetings, or on occasion delivered in face-to-face discussions between individuals. In addition, the mass media is also an indirect source of scientific information, although generally mass media accounts serve to reinforce existing bridging objects rather than present new ones.

Then in a process of consensus building, a "strongly structured local identity" within the bureaucracy is developed. The bridging objects are discussed, argued, and evaluated within the context of other political circumstances and priorities, chiefly at staff meetings, and policy proposals put forth. This begins at the section level (in our case, principally the Air Quality Bureau and the Global Environment Division of the EA). Once consensus is reached, the process is repeated at higher levels if the policy proposals are deemed sufficiently important and relevant. As far as research decisions are concerned, this procedure mainly occurs as part of a coordinated schedule linked to the annual fiscal budget process.

The bridging objects and the policy proposals now associated with them may reach the top of a ministry and then maybe the Council of Ministers for Global Environmental Conservation and the attention of the Prime Minister. During this whole sequence of events, bureaucrats disseminate the bridging objects and policy proposals to LDP and Diet members, and other political actors closely associated with the central government.

In the manner just described, the imported sulfate bridging object and the associated policy proposal (by the EA) to include acid deposition as one of the key problems in GEPRP reached the Council of Ministers for Global Environmental Conservation. The proposal was accepted. However, the existence of the imported sulfate bridging object was not the sole reason why the international acid deposition issue made its speedy climb. Additional reasons (part of the political context of the bridging object) that were attached during the innumerable meetings as the bridging object and policy proposal worked their way up the bureaucratic hierarchy were the fact that by the late 1980s global environmental issues, especially climate change and stratospheric ozone depletion, were garnering great attention in the Western world. Japan became caught up in the tailwind of these issues.[15] In addition, Japan was reeling from repeated attacks on its poor international environmental policy record, and wished to assert a more positive image. Thus, by the time the East Asian acid deposition problem was discovered, the Japanese policymaking world was already ripe for picking up on it. Not a lot of prodding from the scientific community was needed to get it motivated. A window of opportunity existed just as the bridging object appeared, and it facilitated political action, primarily support for research.

But support for Japanese scientists wasn't enough. Even though transboundary transport of acidic pollutants was becoming accepted fact in Japan, this was by no means so in the rest of East Asia. Promotion of research in other countries was necessary if the existence of the problem was to gain widespread acceptance. Thus, Japanese scientists (with support from bureaucrats primarily in the EA, MITI, and MAFF) used the Japanese-created problem-framework and bridging object (existence of regional-scale transboundary transport of air pollutants, in its expanded version) as a vehicle to communicate with non-Japanese scientists and to structure their perceptions on the problem. They accomplished this through activities such as writing journal articles, sponsoring conferences and workshops, and personally meeting other Asian researchers. The first scientific conference to address the transboundary acid deposition problem was the "Acid Rain and Acid Snow in the Circum-Sea of Japan Region" conference held in 1989 in Kanazawa, Japan. The high sulfate deposition data obtained along the Sea of Japan coast prompted this conference. Japanese scientists (and bureaucrats) eventually succeeded in stimulating acid deposition research throughout East Asia and placing the acid deposition issue on East Asia's political agenda. For instance, South Korea and Taiwan both began acid deposition research programs around 1990.

In conclusion, during phase 1 (mid-1980s–1990) the discovery of transboundary air pollution completely transformed not only Japan's acid deposition problem-framework but also its entire notion of sustainability. It confronted the island nation with the certainty that domestic measures were no longer sufficient to achieve sustainability (on acid deposition or many other environmental issues). International cooperation was imperative, and Japan had to step forward to promote such cooperation.

Phase 2: Initiation of International Cooperation (1990–1995)

In 1990 Japan's newly emerging LRT problem-framework still rested on a small and weak "consensus core." Soon, however, scientists expanded this core, elevated its quality, and strengthened the message carried by the existence bridging object. Aircraft measurements begun in January 1991 showed the existence of huge plumes of sulfur dioxide emanating from the continent during the winter months. The first such data was that of Kaneyasu Naoki of MITI's National Institute for Resources and Environment. Kaneyasu's results were not widely known until picked up by the press at an October 1992 scientific meeting. The *Nikei Keizai Shimbun* ran the following headline: "500-Kilometer-Wide Polluted Winter Air Mass from the Continent Hits Western Japan" (8 October 1992, 19; author's translation). Thereafter, the results were even picked up by the foreign press.[16] Other research verified the presence of heavy metal-containing aerosols of continental origin on the Sea of Japan coast (see the appendix for a discussion of this and other data). However, no major negative impacts due to long-range transport were uncovered. Thus, no "impacts" bridging object appeared. Impacts were argued by analogy to the Western experience with acid deposition. More and more it seemed that the big worry was not the present but the future.

Japanese scientists provided a second politically potent bridging object in 1991. Through compilation of detailed emission inventories in East Asia and projection of growth in emissions, they documented for the first time the high emissions in China and their worrisome future trajectories (see the appendix for a discussion of this emissions inventory research). And in the process they created a "trends" bridging object related to future trends of China's acid precursor emissions. In essence, the bridging object stated that China's emissions in the 1980s were very high and growing at an alarming rate. Scientists carried this bridging object to bureaucrats (again, primarily in the EA and MITI). Like the imported sulfate bridging object, it also slowly worked its way up the bureaucratic hierarchy.

The bridging object strengthened bureaucratic incentive to take action. In a way, Japanese policymakers had all the scientific information they needed to act. They knew there was indeed a problem (communicated via the imported sulfate bridging object) and that the likelihood of it getting worse was extremely high (communicated via the China emissions bridging object). Action (policy decisions) came on two fronts, one technological and the other scientific. Japan began to gear up for massive efforts at technological assistance aimed at curbing China's air pollutant emissions, and for intense efforts at scientific cooperation with China and other East Asian nations. In other words, Japan began to proactively pursue regime formation. The centerpiece of Japan's technological initiatives was MITI's Green Aid Plan, announced in 1991 (the acid deposition problem was only one target area in the plan, albeit a significant one as far as China was concerned since a good deal of effort was directed at China's coal-fired power plants).[17] The centerpiece of its scientific initiatives was the EA's EANET (East Asian Acid Deposition Monitoring Network), plans for which were announced in 1992 (discussed later). While these decisions were being made, at no time did the Japanese government make any formal accusations of China's responsibility for the transboundary problem. Environmental diplomacy on the issue by Japanese bureaucrats became active but discreet.

In October 1992 the first Northeast Asian Conference on Environmental Cooperation (*Kan Nihonkai Kankyō Kyōryoku Kaigi*) was held in Niigata, Japan under the auspices of the EA. The EA made no provocative statements about transboundary air pollution during the course of the conference; however, it did urge that "[to prevent] damage of the kind which has been experienced by Europe and North America, it is of prime importance to deal with the problem at [an early] stage in collaboration with the countries concerned" (OECC 1992, 123). China commented: "Air pollutants are characterized by diffusion and long distance transportation. They may cause pollution to adjacent places or neighboring countries. Therefore, strengthening regional cooperation is important" (OECC 1992, 132). The Chairman's Summary proposed that regional cooperation be undertaken to jointly monitor and survey the acid deposition problem (along with other problems such as coastal and inland water pollution and biodiversity loss).

In January 1993, Qu Geping, the first head of China's National Environmental Protection Agency (now State Environmental Protection Administration), admitted the possibility of long-range transport of air pollutants from China to Japan in an interview with the newspaper *Asahi Shimbun* during a visit to Japan (*Asahi Shimbun*, 24 January 1993, 1). The interview was reported in front-page headlines: "Nihon e Ekkyō Osen no Osore" (Fear of Transboundary Pollution to Japan; author's translation). When asked about long-range transport of pollutants, Qu said: "Unavoidable increases in coal production [in China] leading to increased sulphur dioxide emissions may affect [Japan] in the future" (author's translation).

The Second Northeast Asian Conference on Environmental Cooperation was held in Seoul in September 1993 (OECC 1993). In her opening remarks at the conference, the Director-General of the EA, Hironaka Wakako, stated: "It is my sincere hope that the countries in Northeast Asia can create a model of regional environmental cooperation in which preventive measures are taken before serious environmental problems occur. In Europe, it was only after serious cross-boundary environmental damage occurred, *like that caused by acid rain*, that regional cooperative arrangements were achieved. We, the people of Northeast Asia, can still take preventive measures if we cooperate effectively" (Hironaka 1993, 50; emphasis added). Again, the Conference proposed cooperation in regional monitoring of acid deposition. At the third Conference, the EA once again promoted an acid deposition monitoring network in East Asia.

Publication in 1994 of the EA's final report of phase 2 of the Acid Deposition Survey (EA 1994), discussed in detail in the appendix, marks the first "official" Japanese acknowledgment of the existence of long-range transboundary transport of pollutants. The report declares that high wet deposition values of nonseasalt sulfate detected on the western offshore islands of Tsushima, Oki, Sado, and Rishiri during the fall and winter each year of the survey strongly suggested transboundary transport from the mainland, since these islands were little affected by local pollution sources. Japanese newspapers quickly seized on this statement, and in fact it was they who labeled the report the "first official announcement." The *Japan Times*, for instance, stated: "It is the first time the agency has officially suggested industries in China may play a role in acid rain here" (*Japan Times*, 5 July 1994, 3).

The imported sulfur and Chinese emissions bridging objects provided a foundation for communication between Japanese scientists and bureaucrats, and for steering them toward a common regional goal—development of an East Asian-wide regional acid deposition problem-framework (or East Asian LRT problem-framework). Developing such a framework was essential to fostering international cooperation. For the first time in its history, instead of importing acid deposition methodologies, knowledge, and orientation to research, Japan began exporting them for localization in other countries. Japan was not the only actor engaged in this process; however, it was the main one. Another actor, for example, was the World Bank and its Rains-Asia Project.[18] In its newfound leadership role, Japan sponsored numerous international scientific conferences and sponsored international cooperative projects. Illustrative of projects started in the early 1990s are the following.

- *Precipitation Chemistry Monitoring in East Asia Project*, a five-year project initiated by CRIEPI in 1990, and supported in part by MITI. Electric power industry-related research organizations in China, Taiwan, and South Korea joined CRIEPI in conducting joint precipitation chemistry monitoring.
- *Air Pollution and Acid Deposition Impacts to Terrestrial Ecosystems in Chongqing, China Project*, a five-year project sponsored by the Ministry of Education, which started in 1990. Japanese researchers joined with researchers from the Research Center for Eco-Environmental Sciences of the Chinese Academy of Sciences to investigate the impacts of air pollution and acid deposition on the ecology of soils, forests, and lakes in the Chongqing region, the area of most severe acid deposition impact in China.
- *Acid Deposition and East Asian Cultural Properties Project*, a two-year project sponsored by the EA, and begun in 1993. The project studied the effect of air pollution and acid deposition on cultural properties in East Asia. It used identical sample pieces of various metals and

marbles, and measured corrosion rates and chemical composition changes in the samples at twelve locations in China, Japan, and South Korea.[19]

Japan's efforts to build an international problem-framework (and expert community) are reminiscent of Scandinavian efforts in the 1970s. Japan was working hard to create an "acid deposition bandwagon" in East Asia.

By 1995 a rudimentary, low-quality East Asian LRT problem-framework existed. Although Japan dominated the process of creating this framework, and still does to a lesser degree today, its development was not the province of one country. It was a collective effort of most East Asian nations. Since the focus of this book is Japan, the vital and fascinating process of collective construction of the regional problem-framework is only touched upon. Thus, a distinction is made here between Japan's national LRT problem-framework, constructed by Japanese scientists, and the international or East Asian problem-framework collectively constructed by the region as a whole. The two overlap but are not synonymous. It was Japan's national LRT problem-framework that informed its foreign policymaking process.

To give the reader a sense of the East Asian acid deposition problem as it stood circa 1995, thumbnail sketches of the national problem-frameworks of each country, except Japan, are given next. The quality of scientific methodologies suffered in every country except Japan, and perhaps Taiwan. Verification of scientific information was weak, especially as related to transboundary processes; and coverage was spotty. East Asia is a huge region, and only small segments were as yet subject to intensive research. The national problem-frameworks for the seven countries of East Asia collectively added up to a picture of the East Asian acid deposition problem as understood in the mid-1990s.

China[20]

1. The 1979 market-oriented economic reforms initiated under Deng Xiaoping touched off rapid industrial expansion that led not only to phenomenal economic growth but also to skyrocketing emission of air pollutants, some of which were transported across China's borders. The quantity of pollutants transported across its borders, though, paled in

comparison to what stayed behind. Horrendous air quality problems became a symbol of China's rapid development.

2. China's acid deposition problems were severe, and centered in southern China. However, the area receiving acidic precipitation was expanding northward and westward and the acidity of precipitation was increasing. The area receiving precipitation with an annual average pH of less than 5.6 accounted for about 25% of China's total land area in 1993, encompassing almost all of southern China. In 1985 only Chongqing and Guiyang suffered serious acid deposition problems with annual average pHs lower than 4.5.

3. The presence of high concentrations of acid-neutralizing alkaline substances in the atmosphere from the deserts and dry lands in and around China was the only reason China did not have the world's worst acid deposition problem.

4. The source of China's acid deposition problems was primarily coal combustion for energy production. China has the largest coal reserves in the world and is the largest consumer of coal in the world. In the early 1990s, about 75% of China's total primary energy consumption and about 75% of all electric power generation derived from coal. About 50% of coal consumption was concentrated in and around major cities in the early 1990s, and much was burned in small boilers and household stoves for heating and cooking purposes. Almost no desulfurization equipment or other clean coal technologies were employed in either power plants or small boilers.

5. China's SO_2 emissions were expected to double between 1990 and 2020, from about 15 to 30 million tonnes. NO_x emissions were not yet of great concern.

6. Manifest impacts to soils, crops, forests, materials, and properties of cultural importance were all being experienced, especially in southern China. The most severe of these impacts were to forests, crops, and materials. The economic losses from acid deposition impacts seemed to be enormous, especially in the south.

7. Local emission sources contributed most to local acid deposition problems. Exported air pollutants were only a small fraction of total Chinese emissions, on the order of 1 percent.

8. Based on China's knowledge base, Chinese expert consensus was that the domestic acid deposition problem was of crisis proportions. It was therefore imperative to upgrade and expand scientific programs and increase budgets to get a reliable nationwide picture of the true scope of the problem. Also, it was imperative that major control efforts be undertaken. The international dimension of the problem was not serious because only a tiny fraction of China's emissions were exported.

South Korea[21]

1. Precipitation in Korea was only slightly acidic, in the vicinity of pH 5.0. The pHs were high primarily because they were being neutralized by ammonia and alkaline substances (e.g., yellow sands) in the atmosphere. It was estimated that they neutralized about 90% of the acidity of Korea's precipitation in the long-term.

2. Korea's NO_x emissions were increasing, but SO_2 emissions rapidly decreasing.

3. There was little evidence of impacts to inland water bodies, soils, crops, forests, or materials and cultural properties, except in the large urban and industrial areas where ambient air pollutant impacts were likely dominant.

4. Transboundary transport of pollutants was verified through monitoring and long-range transport computer modeling. There were as yet no estimates of total import and export of pollutants.

5. Based on Korea's knowledge base, Korean expert consensus was that the domestic dimension of the acid deposition problem was not serious, but the international dimension was of great concern. Long-range transport of pollutants from China was the chief concern. Thus, science budgets and programs needed to be expanded, and political activity increased.

North Korea[22]

1. Nothing was known of the acid deposition situation in North Korea.

Taiwan[23]

1. The pH of precipitation was low, about 4.5.

2. Neither volcanoes nor the yellow sands phenomenon played a significant role in Taiwan's acid deposition problem.

3. Domestic emissions of SO_2 decreased in the previous decade, but NO_x increased.

4. There was little evidence of damage to ecosystems, crops, materials, or human health.

5. Long-range transport from mainland China was occurring, but the contribution to total acid deposition was unknown. Estimates placed it in the range of 10–50%.

6. Based on Taiwan's knowledge base, Taiwanese expert consensus was that the domestic dimension of the acid deposition problem was not serious, and that the international dimension, while not serious, may become serious especially as southern China continued to industrialize. Continuation of Taiwan's modest national level research project was recommended.

Far Eastern Russia[24]

1. Severe local acid deposition problems existed in far eastern Russia in the vicinity of smelters and industrial complexes.

2. Transboundary import and export of air pollutants seemed minor.

3. Based on Russia's meager knowledge base, Russian expert consensus was that the acid deposition problem in far eastern Russia was not a serious concern as yet, but may become so in the future.

Mongolia[25]

1. Acid deposition was not a problem in Mongolia.

2. Based on Mongolia's scant knowledge base, Mongolian expert consensus was that the country's relevance to the acid deposition phenomenon was as a source region of alkaline substances in the atmosphere from its deserts and drylands, not acidic pollutants.

Although the above compilation of scientific knowledge in East Asia was based on rudimentary and scattered data, several notable features stood out. First of all, there were two large centers of low pH values (southern China and Japan), and one smaller pocket (Taiwan). Each of these areas registered pHs on the order of 4.5. The acidity of precipitation in South Korea, far eastern Russia, and Mongolia was on the order of five to ten times higher, with pHs around 5.0.

Second, emissions of sulfur dioxide seemed to be more or less under control in all countries except China (and possibly far eastern Russia). Emissions of nitrogen oxides were not under control anywhere in East Asia. There were serious NO_x problems in Japan, South Korea, and Taiwan due to the large number of motor vehicles. NO_x was not yet considered a problem in China, far eastern Russia, or Mongolia.

Third, dramatic impacts due to acid deposition were occurring only in southern China. Here the impacts were shocking. Evidence for widespread ecosystem impacts was not found in any other country, with the possible exception of forests in Japan. Acidification of inland water bodies, which is a major source of concern in the West, seemed more or less absent in East Asia.

Fourth, in all countries without exception, ambient air quality issues dominated the acid deposition issue; however, in China the acid deposition issue ran a close second. And fifth, import of acidic pollutants was documented in Japan, South Korea, and Taiwan. However, there was no consensus on the amounts being transported, nor on precise source-receptor relations.

Returning to Japan, by the mid-1990s Japan's LRT problem-framework yielded a clear set of bridging objects, the most important of which addressed potential adverse effects in Japan due to domestic and imported acidic pollutants. Collectively, they can be stated as follows:

Monitoring data demonstrated that precipitation in Japan was acidic at about the same levels as Europe and North America, that there did not seem to be a trend of increasing acidity, and that long-range transport of acidic pollutants from mainland Asia was occurring. Research to date, however, had not revealed any unequivocal, large-scale ecological or nonecological impacts due to acid deposition from either domestic or extraterritorial sources. Impacts to inland water bodies and aquatic organisms did not seem to be a present worry but could become serious in the future. Impacts to forests and forest soil chemistry and ecology were the major concern, but no adverse effects could as yet be unequivocally attributed to domestic or imported pollutants. Rather than documented ecological impacts (as was the case in Europe and North America), the magnitude and rate of increase of precursor emissions in

East Asia and their future ecological and nonecological impacts was the greatest cause for concern.

The impacts bridging objects within this set, while not easily lending themselves to dire predictions of impending doom, were still a powerful motivating force in Japan's international environmental diplomacy because they raised red flags about potential future impacts.

Phase 3: Establishment of EANET (1995–2001)

A third phase in the evolution of the East Asian transboundary air pollution issue was evident by 1995. It was defined by the efforts to establish EANET, which became the focal point in the regime formation process.[26] MITI's Green Aid Plan, the other foreign policy centerpiece during the second phrase, suffered severe problems in China. The high cost of Japan's technology and lack of personnel with adequate technical skills to operate and maintain it, for instance, dampened enthusiasm for the plan. Thus, it more or less dropped out of the picture as a driving force in acid deposition foreign policy during the third phase.[27]

China still downplayed the seriousness of the international dimension of the problem. Japan, while still the most proactive country, was no longer alone. South Korea was active scientifically and politically. Taiwan also became active, hosting for example an "International Conference on Acid Deposition in East Asia" in May 1996 (International Association for Meteorology and Atmospheric Sciences 1996).

The method by which EANET come to be the centerpiece of Japan's acid deposition foreign policy reveals again the scientist–bureaucrat interaction highlighted earlier. The idea for a regional monitoring network was put forth by scientists familiar with the scientific approach to the European regional acid deposition problem, where EMEP got off the ground about a decade after the discovery of the problem. By 1991 the Asian monitoring network idea was associated with the two main science-to-policy bridging objects (imported sulfate and China emissions) and, as a policy proposal, had worked its way up the EA bureaucratic hierarchy. It was accepted by the top EA leadership as a central policy objective and was included in the EA's budget request for 1992.

Besides the impetus given by the bridging objects, this decision gained support because of the opportunity it provided for the EA to establish itself in regional environmental issues, the potential it gave the agency to operate on an international environmental issue more or less independent of MITI, and the EA's historical scientific momentum (i.e., experience) with monitoring networks. These factors constitute the context for the bridging objects and were some of the key reasons given in the consensus-building process within the EA to push forward with the proposal.

The next hurdles were to gain approval from the Ministry of Foreign Affairs to operate internationally in this issue area and from the Ministry of Finance for funds for the 1992 fiscal year to pursue the network. The EA gained approval from both primarily on the basis of its experience with monitoring networks, the precedent established by Europe, the importance of the issue to Japan as spelled out by the bridging objects, and the interest of other Asian countries in the idea. Eventually, too, the EA succeeded in convincing the Council (and Prime Minister who headed the Council) to support EANET after proactive dissemination of the LRT bridging objects, judicious lobbying of other ministries, assiduous cultivation of interest by other Asian countries in the monitoring network, and careful reminders of its privileged position in regards to acid deposition monitoring.

The EA's first announcement that it would seek to establish an acid deposition monitoring network in East Asia occurred soon after the United Nations Conference on Environment and Development (UNCED, or "Earth Summit") in August 1992.[28] The first step was to be a series of "expert meetings" of scientists and administrators from the region. The "First Expert Meeting on Acid Precipitation Monitoring Network in East Asia [sic]" was held in October 1993 in Toyama, Japan (EA 1993). More than forty scientific experts and administrative officials from ten countries and three international organizations attended. The participating countries included China, Indonesia, Japan, South Korea, Malaysia, Mongolia, Philippines, Singapore, Russia, and Thailand. The meeting was the first to bring together both scientists and policymakers. It was primarily devoted to ascertaining the state of acid deposition monitoring and research programs, and known impacts, in the participating countries. There was apprehension initially that the differing interests of the countries would lead to clashes, but such fears proved groundless.

The second Expert Meeting was held in March 1995 in Tokyo (EA 1995a).[29] It was agreed that in light of the vastly differing social, economic, political, and geographic conditions of each country a "step-by-step" approach would be taken to establishing the network. A third meeting was held in November 1995 in Niigata (EA 1995b). This meeting finalized the "Conceptual Design of an Acid Deposition Monitoring Network in East Asia," agreed to establish an interim acid deposition monitoring network center in Japan, and urged that the network be formally established "as soon as practical, but no later than the year 2000." A fourth meeting was held in February 1997. Attendees adopted technical manuals, and proposed a work plan and timetable for realizing the network. The achievements of the four Expert Meetings between 1992–1997 are summarized in EA (1997a), and the Guidelines and Technical Manuals for the network are contained in EA (1997b). The next big task was to formally propose establishment of the network.

In June 1997 at the United Nations General Assembly Special Session on Environment and Development held in New York to honor the fifth anniversary of the "Earth Summit," Prime Minister Ryutaro Hashimoto proposed, as part of Japan's so-called "Initiatives for Sustainable Development," establishment of EANET. He signaled Japan's intent to make EANET a central pillar in its efforts to promote sustainability in the East Asian region. EANET "will be dedicated to creating common understanding of the status of the acid deposition problem among the countries in the region through implementation of acid deposition monitoring by each country, central compilation, analysis and evaluation of such monitoring data, and periodic publication of reports. Such common understanding will become the scientific basis for taking further steps to tackle the problem, such as cooperative measures to control precursor emissions" (Interim Secretariat of EANET 1998a, 18). In the terminology of this book, the Prime Minister was proposing that EANET itself become a bridging object among the nations of East Asia. Hashimoto then offered to convene an intergovernmental meeting to establish EANET.

The First Intergovernmental Meeting on EANET (March 1998) was preceded by two meetings of the Working Group on EANET (November 1997 and March 1998).[30] Although formal establishment was put off until a second meeting, the first meeting designated the Acid Deposition and Oxidant Research Center (ADORC) in Niigata, Japan as the interim network center. A Third Meeting of the Working Group and a

First Meeting of the Interim Scientific Advisory Group met back-to-back in October 1998.[31] The big news at the October meetings was that China, which had resisted formal participation in the network, announced that it would become a member of EANET.

EANET finally came into being at the Second Intergovernmental Meeting held in Niigata, Japan in October 2000.[32] Ten countries joined the network—China, Indonesia, Japan, Malaysia, Mongolia, Philippines, South Korea, Russia, Thailand, and Vietnam. The Joint Announcement declared: "Recognizing that East Asia is facing increasing risks of problems related to acid deposition, . . . [participating countries] will cooperatively start the activities of EANET on a regular basis from January 2001, in a transparent manner to achieve the following objectives: (1) to create a common understanding of the state of acid deposition problems in East Asia; (2) to provide useful inputs for decisionmaking at local, national and regional levels aimed at preventing or reducing adverse impacts on the environment caused by acid deposition; and (3) to contribute to cooperation on the issues related to acid deposition among the participating countries."[33] Institutional arrangements include Intergovernmental Meetings of Participating Countries (the main decision-making body), a Scientific Advisory Committee, a Secretariat hosted by UNEP, a Network Center (designated as ADORC in Niigata), and National Centers in each country.

After eight years of intense effort on the part of the Environment Agency of Japan, Asia's first regional-scale, scientific monitoring network (of any kind) was born. During these eight years, the EA picked up virtually all expenses associated with launching the network. By the time EANET formally began operation in January 2001, some fifteen years had elapsed from the discovery of the East Asian transboundary problem. This comparative slowness is indicative of the uphill battle in East Asia to develop regional cooperative institutions, environmental or otherwise.

Regime Formation and Regional Sustainability

The discovery of the transboundary air pollution issue forced Japan to acknowledge that international cooperation and regime formation was

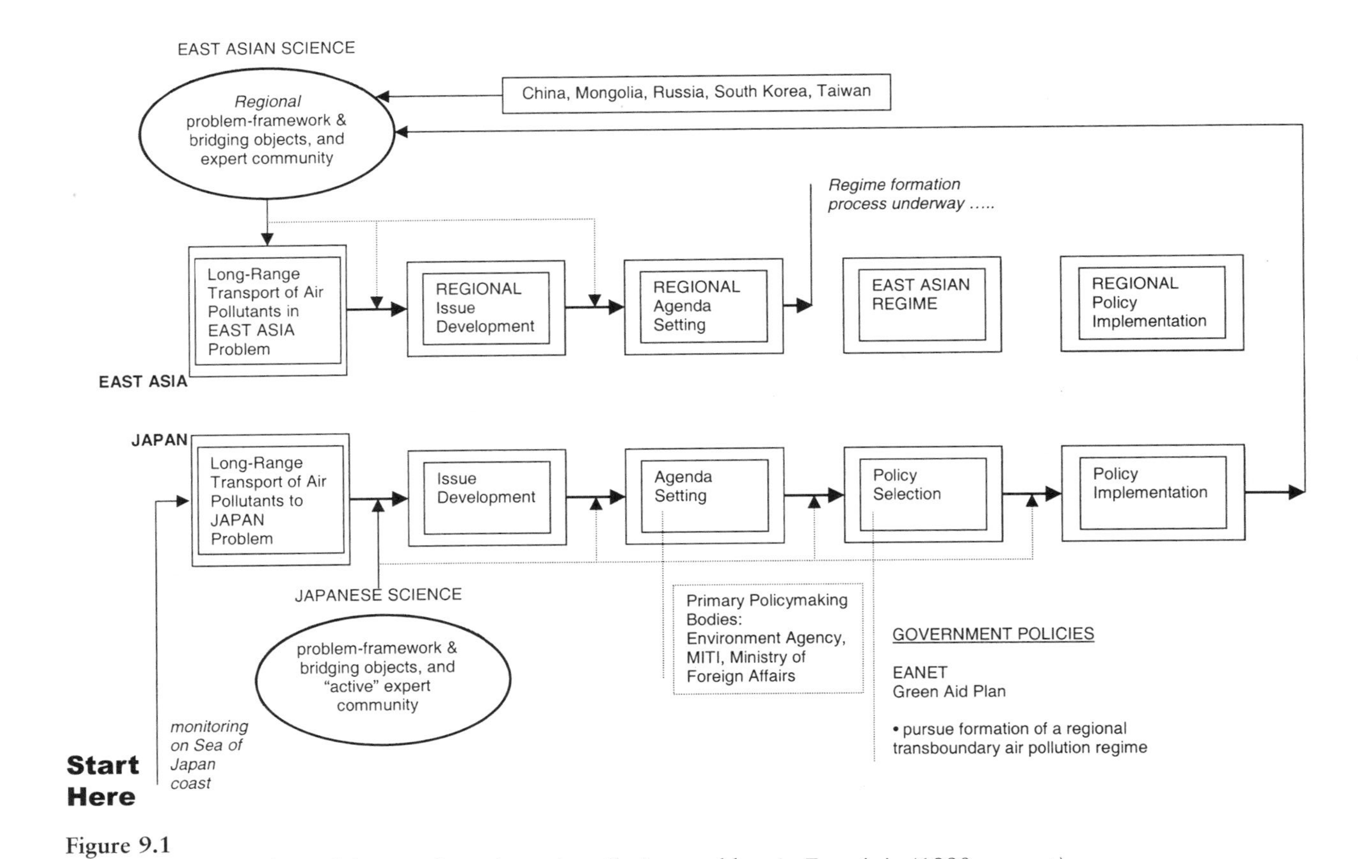

Figure 9.1
Science-policy interface of the transboundary air pollution problem in East Asia (1990–present)

imperative to achieve acid deposition sustainability. Japan soon developed a two-pronged approach—pursuit of technology development, transfer, and assistance (e.g., MITI's Green Aid Plan), and pursuit of an international acid deposition scientific network (i.e., EANET). The technological prong lost some of its steam; however, the scientific prong—EANET—has succeeded nicely and is now seen as the first step toward a signed treaty (like those already achieved in Europe and North America) to address problems related to the transport of air pollutants across national borders in East Asia. In essence, regime formation is the solution path spelled out by Japan's LRT problem-framework.

The problem-framework, however, does not yet contain specific sustainability criteria. Thus, the path is incomplete. The knowledge base is insufficient to determine whether the present rate of acid deposition in Japan is sustainable. However, Europe's critical loads approach is receiving significant attention. Currently there are two groups engaged in such research, the CRIEPI Acidic Deposition Project and a Critical Loads Research Team under the auspices of the EA.[34] Both groups seek to develop critical load maps as in Europe, and both CRIEPI and the EA aim to develop critical loads-based integrated assessment models similar to the RAINS model used in Europe. It is too early to tell, though, whether these tools will become central to acid deposition policy in Japan, or whether some other method will supersede them; however, it is clear that sustainability criteria will best be developed under the umbrella of an East Asian cooperative scientific structure (versus independent national level research).

Regime formation, therefore, is vitally important to Japan and East Asia. It is likely to be a long, hard process, but the effort itself is already pioneering approaches that will inform knowledge and provoke action on other regional environmental problems. A transboundary air pollution treaty would be the first such treaty in East Asia, and would provide precedent for other issues. Japan's expert community and problem-framework are blazing a cognitive trail that will sooner or later, I believe, lead to a regional treaty on transboundary air pollution. Their role in this process is illustrated in figure 9.1.

10

Sustainable Rain: Knowledge and Action

The Japanese people began losing their centuries of pristine air and natural rain soon after the Meiji Restoration of 1868. The air and rain were first contaminated by acidic smoke from copper smelters. They were further polluted by acidic industrial and residential emissions in growing urban areas in the prewar era. Air and rain continued to be polluted by even more acidic emissions during the reconstruction period after the war. In the 1970s the gentle mist and drizzle of the rainy season turned caustic in the Tokyo area from moist air pollutants.

Thereafter, acid rain was found to be falling nationwide and soon even school children began measuring rainfall pH. And now acidic pollutants travel long distances from far shores to poison the air and rain of Japan. For over a hundred years Japanese scientists and policymakers have struggled to understand acidification phenomena and to cleanse the atmosphere of unnatural acidic substances. Much can be learned from this history. There are two dimensions to what can be learned (and to the analysis in this book). The first relates to Japan's acid deposition history in and of itself, and the second to the interaction between science and politics in general, independent of Japan.

As portrayed in this book, problem-framework and bridging object are master concepts for understanding the role of ideas at the science-policy interface. An environmental problem-framework is a geographically and historically specific "package" of expert knowledge, methods, and orientations related to an environmental focus-problem. Thus, it is a codified body of ideas constructed by and used by experts to provide cognitive structure (a framework) for researching and understanding a problem such as acid deposition. A problem-framework is analogous to Thomas Kuhn's concept of a paradigm. It is not, however, necessarily related to "the structure of scientific revolutions" at the cutting edge of science. It more generally relates to smaller scale organization of scientific knowledge and practice, and, in my usage, is associated with politics. Just as Kuhn's paradigms serve a patterning function for science during "normal" periods of scientific inquiry, problem-frameworks serve a patterning function for scientific knowledge related to a focus-problem during a given historical period and for the given political jurisdiction. Problem-frameworks are the product of an identifiable group of experts, referred to in this book as an expert community. Expert communities are legitimized builders of problem-frameworks.

There is nothing inherent in the logic of science that inevitably leads to its extension into policymaking. A seldom acknowledged conclusion about the science-policy interface that careful historical tracking of scientific ideas in Japan's acid deposition history clearly reveals is the vast flow and alteration of scientific knowledge that prefigures any obvious influence on politics. A starting point for the political influence of science is initial construction of a localized problem-framework relevant to a given policymaking jurisdiction. Localization of scientific knowledge is essential to politics. It is well and good that Sweden and Canada have acid deposition problems, but if it cannot be shown that the same exists in Japan, then Japanese policymakers have no incentive to pay attention to the problem.

Local construction of problem-frameworks is premised on the existence of scientific concepts and methods with universal attributes that allow them to circulate globally in the scientific world. There must be certain qualities to acid deposition scientific knowledge, for instance, that allow it to be internationally transmitted to Japan and transformed upon arrival. Transmission is accomplished by means of science-to-science bridging objects that contain "universal ideas" communicated from scientist to scientist. As we have seen, Japanese scientists expended phenomenal effort over many decades adapting "universal" acid deposition ideas developed in the West to Japan. We were not much concerned about the initial creation of acidification ideas because most of them originated in Europe and North America, not Japan.

The localization process is premised in one way or another on political support for science (in this sense, science and policy "co-construct" each other). In the case of Japan's acid deposition history, the nature of this support changed. During Japan's first three acid deposition periods (1868–1974), acid deposition science advanced due to general government support for science. In contrast, during the last three periods (1974–present), there has been targeted support for acid deposition research.

Problem-frameworks are not static; they undergo change. When changes in the body of ideas are such that a fundamental shift in knowledge, methods and/or orientation occurs, the problem-framework is said, in the language of this book, to undergo an epistemic shift. Each of Japan's six acid deposition periods is characterized by a unique problem-framework. The shift in the problem-framework occurred as follows: copper smelter smoke (period 1) → acidification of urban precipitation (period 2) → general acidification of precipitation (period 3) → moist air pollution (period 4) → nationwide acid deposition and its ecological impacts (period 5) → import of acidic pollutants transported from mainland Asia (period 6).

The shifts were for the most part slow, occurring over many years. Portions of each problem-framework were passed on, but the process was not necessarily accumulative. Thus, Japan's acid deposition history is not a smooth and linear buildup of scientific knowledge. Epistemic shifts were also generally accompanied by shifts in the composition and character of the expert community, although these shifts were less dramatic than shifts in the problem-framework.

What causes epistemic shifts? Japan's acid deposition history is primarily a story of acid deposition-related ideas imported from the West that, once localized to Japan, prompted epistemic shifts. Thus, one answer is that import of foreign ideas causes epistemic shifts. But this is not the complete answer. Besides knowledge-related international forces, domestic forces were also at work. And forces completely outside the domain of science were clearly operative. Between the first and second periods, besides import of Western ideas related to polluted urban atmospheres, worsening urban air quality and ripening interest in indigenous research were factors inducing the shift.

Between the second and third periods, horrendous air quality nationwide and the curiosity of a few scientists in acidification phenomena were factors. Between the third and fourth periods, the sudden appearance of moist air pollution, not import of ideas, triggered the shift. Between the fourth and fifth periods, low pH readings throughout the country and an active expert community, combined with acid rain news from the West, helped facilitate introduction of the acid rain problem-framework of the West. And between the fifth and sixth periods, growing awareness of international environmental problems around the world helps explain the shift to a problem-framework that emphasized the international dimension of Japan's acid deposition problem.

Thus, no single mechanism caused Japan's epistemic shifts. However, all were fundamentally premised to one degree or another on (1) indigenous development of domestic scientific capacity, and (2) vigorous international trade in scientific ideas. These are not independent processes. Indeed, they were cofactors in precipitating the shifts. We observed throughout its acid deposition history Japan's proverbial power to learn from abroad, digest in a process of domestic internalization, and then innovate on what it absorbed. Learning from abroad helped establish its domestic scientific capability (as was done in the Meiji period). And conversely, establishing scientific capability provided powerful incentive to learn from abroad (as was done in the aftermath of Japan's experience with moist air pollution). These processes point to two important elements to the flow of ideas between scientists—an international flow (primarily from the West to Japan) and an *intra*national flow (primarily from national-level, elite scientific institutions to local scientific institutions). The general pattern throughout most of Japan's acid deposition history was one of Western scientific ideas stimulating Japanese scientists, principally those in elite scientific institutions, who then in turn stimulated scientists at local scientific institutions. The end result was a deep and solid nationwide foundation from which science could inform policymaking.

During the sixth (and present) period, Japan is finally running into the limits of its traditional, one-way relationship between international trade in ideas and domestic scientific capacity building. Today Japan is creating knowledge at the frontiers of acid deposition science, and is engaged less in one-way learning from abroad and more in coequal exchange. In addition, Japan has become teacher to less developed countries as they deal with their emerging acid deposition problems.

An epistemic shift in problem-framework does not necessarily entail a corresponding shift in the influence of science on policymaking. Thus, a shift in scientific viewpoint alone is insufficient to explain policy action. There were only two periods during Japan's acid deposition history when problem-frameworks and expert communities significantly influenced and structured domestic and/or foreign policy, and when a temporarily stable definition of an acidification focus-problem was adopted by both the scientific and policy worlds. In period 1, science and policy interacted but the problem-framework was too ill-defined and the expert community too amorphous to be of great influence. During periods 2, 3, and 4, science did not influence policy. In period 5, a nationwide, ecologically oriented acid deposition problem-framework structured both scientific investigation and research policy. In period 6, the LRT problem-framework now structures research and policy, and the approach to regime formation.

How do we explain the influence of science on policy in periods 5 and 6? Here is where the second master concept, the science-to-policy bridging object, enters. Problem-frameworks for technical, subtle problems such as acid deposition basically reside within the expert world of the expert community. They are not readily accessible to the non initiated. Thus, a vehicle is needed to carry the technical ideas of the problem-framework into the political world. Science-to-policy bridging objects are such a vehicle. They are bridges between knowledge and action. They are lay-translated elements of the problem-framework created and carried by experts into the political realm where they serve a patterning function for policymaking, a basis from which political actors can determine their interests relative to the focus-problem. Different political actors will likely interpret bridging objects differently. Thus, by themselves bridging objects do not determine policy; however, their character and context help explain their interpretation by political actors.

The political actor that received greatest attention in this book was scientists. Their political influence was described primarily in terms expert community activism. And activism was described largely in terms of the ability of members of the expert community to proactively create and interpret politically potent science-to-policy bridging objects to policymakers. Essentially these experts directly or indirectly were saying to policymakers "because of the existence of such-and-such a bridging object you should think in terms of doing so-and-so."

During periods 5 and 6, I argued that the two main political actors were scientists and bureaucrats. The policy-active within the acid deposition expert community (generally from the ranks of scientists at elite scientific institutions) synthesized, summarized, and lay-translated for consumption by bureaucrats the problem-frameworks of these periods via a set of politically relevant bridging objects. Their fora of interaction included progress and fiscal year-end reports and presentations, conferences, occasional face-to-face discussions between individuals, and interaction with the mass media.

From mid-level bureaucrats the bridging objects generally worked their way up the political hierarchy, and came to be associated with various policy proposals. Japanese scientists, for the most part, are very "Confucian" (circumspect, quiet, cautious, and respectful) in their interaction with other political actors. It is noteworthy that a policy-activist group never materialized during periods 5 and 6. Fora such as "non-sanctioned" press conferences, signed petitions, protest meetings, lobbying in the Diet, etc., have not, to date, been part of the acid deposition expert community's repertoire of influence. It is also noteworthy that few knowledge brokers exist to aid in the transmission of acid deposition-related scientific knowledge to policymakers. This strengthens the scientist-bureaucrat linkage and stands in sharp contrast to most Western nations where, for example, large and well-funded NGOs are valuable sources of lay-translated scientific knowledge for policymakers.

Returning to bridging objects, how exactly do bridging objects influence policy? Their character (type and density) and context are important. In this study, the following types of science-to-policy bridging objects were most important: existence, extent-intensity or trend, and impact or risk assessment. Density refers to the number of "active" bridging objects floating in the political world (lots or few). During periods 5 and 6, the bridging objects of greatest influence on Japanese bureaucrats and other political actors were those that highlighted future and unknown impacts. In period 5 the "low pH precipitation throughout the country" bridging object was the basis for arguing that the types of ecological impacts uncovered in the West (especially damage to freshwater and forest resources) could be occurring in Japan; thus, research programs needed to be established and maintained. The main domestic policy outcome was establishment of the Acid Rain Survey. In period 6 a greater number (density) of bridging objects—in particular, the "imported sulfur" and "China emissions" bridging objects—were used to argue the possibility of irreversible ecological and nonecological impacts from imported pollutants. The main foreign policy outcomes were EA's EANET and MITI's Green Aid Plan, of which EANET was the more successful. In addition, the bridging objects motivated Japan's dogged pursuit of regime formation.

Context was also critical in determining the influence of science on policy in periods 5 and 6. Two elements of context were emphasized in this study—political context and cultural context. In period 5 the domestic political context of the EA at the time (e.g., free time to devote to a new project) led to the birth of the Acid Rain Survey in 1983. In period 6 the international political context (e.g., Japan's poor environmental reputation) was again critical in supplementing science-related influences. The second element of context—culture—shaped science's interaction with politics in at least two important ways. It conditioned the practice (e.g., Japanese scientists' behavior toward policymakers) and acted as a filter through which bridging objects passed as they entered the political world (e.g., Japan's culture of rain and culture of wood). Understanding political and cultural context is necessary to understand the influence of science on political actors.

In conclusion, as illustrated by Japan's acid deposition history, the influence of science on politics is organic and complex. I used comparative analysis of the six acid deposition periods between 1868 and 2001, examining what was present and what missing in each period, to single out key dimensions of this influence, and deduced the following: Science influences politics by means of a problem-framework that, when lay-translated via an array of science-to-policy bridging objects, may be adopted by the scientific, policy, and other social worlds involved to construct a temporarily stable definition of the focus-problem and a thriving line of research and policy innovation (reminiscent of Fujimura's "standard package;" see chapter 2). Specific elements of influence include: (1) the construction of a problem-framework localized to a political jurisdiction by an expert community, (2) the creation of an array of science-to-policy bridging objects, (3) the activism of members of the expert community in carrying and interpreting the bridging objects to the political world, (4) the political and cultural contexts of the objects, and (5) the historical scientific momentum related to the problem.

This book is ultimately about human understanding of large-scale, long-term natural/social phenomena as determined by environmental science, and human management of large-scale, long-term nature/society interactions as determined by national and international environmental politics. It addresses the new generation of "invisible" environmental problems, such as acid deposition, whose effects are indirect, long-term, low-level, slowly accumulating, and widespread. These highly technical problems present a monumental challenge to achieving sustainability. If nothing else, the history of the acid deposition problem in Japan tells us that the path to sustainable societies will not be an easy one. While the path may be rough it is not impassable.

An essential institutional development necessary to walk the path is a positive and dynamic complementarity between science-making and policy-making in service to sustainability. Scientists and scientific knowledge are not stand-alone elements of influence in the policymaking process on complex environmental issues such as acid deposition but integral components. Environmental science and politics are yin and yang; they co-construct each other. Casting them in terms of the millennial-old Oriental concept of mutual interaction between complementary (and coequal) opposites does more than merely state the obvious that science and politics need each other on complex environmental issues. Perhaps the most succinct expression of their true relationship is the famous aphorism of the famous fifteenth-century Chinese official and Neo-Confucian scholar, Wang Yang-ming: "knowledge is the beginning of action; action the completion of knowledge." Knowledge [science] and action [policy] are a unity—in true knowing there is effective action, and in action the knowing is already included.

Japan's acid deposition history instructs us, however, that there are no simple formulas expressing this "unity" in relation to sustainability. Overall, a long-term pattern of increasing intimacy and feedback (increasing complementarity) between science and policymaking seems apparent and seems to be tied to the increasing scale and complexity of the acid deposition focal-problem. As the scale of the problem expanded from local (copper smelters) to national (urban/industrial emissions) to international (extraterritorial sources), and as it moved from simple (one source type) to complex (multiple source types, domestic and international), the corresponding relationship between science and policymaking became tighter with growing mutual feedback (if not influence). This observation may simply echo the obvious fact that between 1900 and 2000 in Japan, as elsewhere in the world, the human impact on the environment increased such as to virtually demand greater interaction between science and policymaking.

My dissection of Japan's acid deposition history points to a few suggestions for enhancing the complementarity between science and policy. First, the relationship can be enhanced by strengthening the ability of scientists throughout the world to construct localized problem-frameworks. In other words, domestic scientific capacity must be improved worldwide to correct its asymmetrical concentration in Europe, North America, and Japan.

Second, the relationship can also be enhanced by cultivating the art of fashioning bridging objects with the purpose of developing a high density of material and conceptual links between the scientific and political worlds. The bridging object notion is deceptively simple but fundamental to communication between these disparate worlds. Bridging objects are not necessarily easy to create. Often because of time, energy, and funding pressures, scientists do not put the effort into crafting effective communication tools such as bridging objects. Perhaps popularizing a term like "bridging object" may make scientists more aware of what they need to create and why.

And finally, the relationship between science and policy can be enhanced by focusing attention on solution paths containing ecologically based sustainability criteria that attempt to interpret nature's ground rules. While this is a tall order (the precautionary principle is a statement of its difficulty), insisting on this orientation clearly emphasizes the truism that a healthy human society is fundamentally premised on a healthy natural world.

There is a tendency in Japanese society (and the global community) to place social sustainability (the maintenance of human health and livelihood) before ecological sustainability (the maintenance of essential ecosystem functions). Japan's acid deposition history allows us to observe the changing tie between them, and suggests that the shift from social sustainability to "ecological + social" sustainability involves an arduous shift in perception and values. We saw several times in Japan's acid deposition history that when environmental problems pushed the limits of social sustainability, they drew action. We saw this in the Ashio Copper Mine case, in the domestic environmental revolution of the late 1960s and early 1970s, and in the moist air pollution incidents. However, when environmental problems push the limits of ecological sustainability, they may not draw action. Why?

One reason is that it may not be easy to detect if a situation is ecologically unsustainable. To detect such a condition often demands sophisticated environmental science. Even Japan's highly developed acid deposition science is unable at this point in time to state definitely whether the present deposition of acidic substances in Japan is unsustainable. Thus, in the terminology of this book, the inability so far to develop a solution path containing scientifically based sustainability criteria relative to acid deposition hinders political commitment to ecological sustainability. Europe's solution path based on the critical loads concept and RAINS integrated assessment model is more successful. Serious work in Japan on critical loads and an integrated assessment model did not start until the mid-1990s. It needs sustained commitment. Thus, while the solution path of the LRT problem-framework defines acid deposition as an international collective action problem, it isn't sufficiently mature to define sustainability criteria to guide what specific actions are necessary to get the region in line with ecosystems' ability to absorb acidic inputs.

The ultimate effect of a solution path, however, may not be to specify nature's ground rules per se but to change human perceptions and values. Further epistemic shifts related to the problem of acidification of the atmosphere may trigger further perception and value shifts in Japan. Recent evidence points to intercontinental transport of acidic and other reactive air pollutants. Japanese scientists have helped produce this evidence. Thus, acid deposition may be becoming a hemispheric, if not global, problem. Perhaps the expanded scale and complexity of the problem, combined with new evidence of adverse effects, will induce the next epistemic shift that will in turn prod a perception and value shift that moves Japan (and the global community) toward a treaty to protect the global atmosphere and toward sustainability relative to acid deposition and other environmental problems.

Appendix
Japan's Present-Day Acid Deposition Science

Japan is one of the world's premiere centers of acid deposition-related scientific research, and possesses a sophisticated acid deposition-related problem-framework developed by a large and active expert community. This appendix introduces Japan's present problem-framework related to the long-range transport of acidic air pollutants as it stood circa 2000. The framework is referred to as the long-range transport or LRT problem-framework. In the first section, the major acid deposition-related research projects and programs in Japan are explained. It is under their umbrella that most research on the problem is conducted. In the second section, the scientific knowledge accumulated as a result of these diverse projects and programs is outlined. Japan's monitoring networks, emission inventories, atmospheric computer modeling, and ecological and nonecological impacts research are discussed. In addition, some of the methods used to construct the knowledge base are highlighted.

Between the mid-1980s and mid-1990s Japan was more or less preoccupied with absorbing the acid deposition-related LRT problem-frameworks of Europe and North America. However, since the mid-1990s it has increasingly established an indigenous pool of knowledge based on indigenous methodologies specifically suited to the Japanese and East Asian context, and hence, rapidly developed an indigenous LRT problem-framework.[1] The three research projects contributing most to this process are: (1) the EA's Acid Deposition Survey, (2) CRIEPI's Acidic Deposition Project, and (3) the EA's Global Environmental Research Fund (GERF). These projects are discussed next. In section 2, after a description of the knowledge bases for component areas of the problem-framework (i.e., monitoring networks, emission inventories, atmospheric computer modeling, and ecological and nonecological impacts), the bridging objects to emerge from the knowledge base are tallied and their quality evaluated.

Major Scientific Research Programs

There is currently a vast amount of acid deposition research being conducted in Japan. The discovery of long-range transport of air pollutants in the East Asian region not only resulted in an epistemic shift to a framework dominated by the long-range transport issue, but also precipitated a veritable boom in acid deposition programs and funding. Research is conducted by a wide range of organizations, including national and local government environmental research institutes, universities, and industrial research laboratories. There are five major acid deposition-related research programs in Japan: (1) the Global Environmental Protection Research Plan (GEPRP), (2) the Global Environmental Research Fund (GERF), (3) the Acid Deposition Survey, (4) CRIEPI's Acidic Deposition Project, and (5) the Association of Local Government Environmental Research Institutes' acid deposition projects. Each is briefly described here.

Acid Deposition Research within GEPRP

The Global Environmental Protection Research Plan (GEPRP), or in Japanese, *Chikyū Kankyō Hozen Chōsa Kenkyū nado Sōgō Suishin Keikaku* (literally, Global Environmental Protection Surveys, Research, etc. Comprehensive Promotion Plan) was established in 1990 and serves as the central government's master plan for coordinating government activities related to conservation and protection of the global environment—activities such as research, surveys, observation programs, monitoring networks, technology development, and so forth.[2] It is overseen by the Council of Ministers for Global Environmental Conservation, a cabinet-level council of all ministers whose jurisdiction involves some aspect of the global environment. (The Council's origin and function was explained in chapter 9.) At present there are eight core research areas within GEPRP: climate change, depletion of the ozone layer, acid deposition, marine pollution, biodiversity loss, tropical deforestation, desertification, and "other."

GEPRP is the umbrella under which most acid deposition-related research in Japan occurs. Research programs and projects begun before 1990, such as the Acid Deposition Survey, have been folded into the plan. Funding for acid deposition-related research and technological development is only a small fraction of the total GEPRP budget, on the order of 0.3% in the mid-1990s. The great majority of the budget is devoted to climate change. Within the acid deposition portion, technology development—especially that conducted by MITI—dominates, constituting roughly 95% of the total budget. (Technology development by MITI likewise dominates the other research areas in GEPRP.) The total GEPRP budget increased from about U.S.$4.4 billion in 1990 to about U.S.$5.8 billion in 1995, and the acid deposition-related budget from about U.S.$4.5 million to U.S.$14.4 million. (Note: The U.S. NAPAP was funded at a level of about $60 million per year for ten years between 1980 and 1990.)

Acid Deposition Research within GERF

Within GEPRP, the Environment Agency's Global Environmental Research Fund (GERF), or in Japanese, *Chikyū Kankyō Kenkyū Sōgō Suishin Hi* (Comprehensive Promotion of Global Environmental Research Fund) was also established in 1990. One subprogram of GERF is the Acid Deposition Research Program. It is the focal point of most basic research (as opposed to survey work) on acid deposition. GERF channels funds to a wide range of research projects. There were four main topic areas within the Acid Deposition Research Program of GERF by the mid-1990s:

1. Distribution and behavior of acidic substances in East Asia
2. Impact of acid deposition on natural ecosystems
3. Critical loads research
4. Development of a policy-oriented integrated assessment model of acidic substances in East Asia.

One of the main objectives of the last topic area is to develop strategies for reducing emissions of acidic substances in China. All topic areas, and subtopics within them, have changed significantly over the years, reflecting alterations in acid deposition thinking in Japan. For instance, critical loads research didn't appear until 1993.

Acid Deposition Survey

Besides GERF, the Environment Agency also coordinates the Acid Deposition Survey (*Sanseiu Chōsa*), the most comprehensive survey of the state of the acid deposition problem in Japan. Unlike GERF research, the survey does not engage in fundamental research; rather, it seeks to document the state of precipitation chemistry in Japan and to gather data on distributions of forest damage, soil susceptibility to acidification, and so forth. Similar to GERF, a wide array of ministries and agencies are involved in the various subprograms within the survey. As described in chapter 8, the survey began in 1983 and is currently in its fourth five-year phase.

The results of phase 1 of the Acid Deposition Survey were discussed earlier. The results of phase 2 of the survey were published in Japanese (EA 1994) and English (EA 1995c).[3] The results of phase 3 were published in Japanese (EA 1999). Phase 2 and phase 3 results are discussed in the next section. The Survey has four major project areas: atmospheric system, inland waters systems, soil and vegetation systems, and integrated impact analysis. The subproject areas for phase 2 are listed in table A.1.

CRIEPI Acidic Deposition Project

The Central Research Institute of the Electric Power Industry (CRIEPI), the research arm of the electric power industry, began a comprehensive acid deposition research project in 1987.[4] It is the most tightly coordinated and highly integrated acid deposition program in Japan. CRIEPI began studies of the environmental impact of electric power plants in Japan in the 1970s, of which one aspect was precipitation chemistry. This led to the Inland Sea Acid Deposition Project conducted in 1984, and described in chapter 8. Based on these and other research experiences, and increasing interest in the acid deposition problem in Japan, CRIEPI established the Acidic Deposition Project in 1987.

The overall goal of the project is to quantify and assess the effect of emissions from electric power plants in Japan on Japan's acid deposition problem. At first the focus was strictly domestic; however, by 1990 the scope expanded to include East Asia. The project was classified as one of two problem areas within CRIEPI's long-term research strategy on

Table A.1
Project areas within the Acid Deposition Survey: Phase 2 (1988–1993)

Project Area	Subprojects
Atmospheric system	• Present state of acid deposition in Japan • Development of long-range transport computer model
Inland waters systems	• Acid snow survey • Groundwater impacts survey • Rivers, lakes, and wetlands impacts survey • Water quality of natural lakes and wetlands
Soil and vegetation systems	• Soil monitoring • Forecasting soil impacts • Priority survey of effects of acid deposition on soil and vegetation in forest decline zones • Monitoring of ecosystem impacts
Integrated impact analysis	• Integrated impact assessment pilot project • Effect of acid fog on forest ecosystems

"global environmental problems." (The other problem area is climate change.) The CRIEPI project is divided into four main research areas: dry and wet deposition monitoring and short-term ecological surveys; long-range transport computer modeling, which includes compilation of emission inventories; research on the long-term effects of acid deposition on inland waters, soils, and forests; and comprehensive assessment of the acid deposition problem in Japan and East Asia. CRIEPI aims to complete a critical loads-based integrated assessment model similar to that planned by the Environment Agency's GERF program. Phase 1 of the project lasted from 1987 to 1993, phase 2 from 1993 to 1997, and phase 3 from 1997 to 2001.

The Federation of Electric Power Utilities (FEPU) and MITI fund the Acidic Deposition Project. From 1987 to 1993 the budget was about U.S.$2.2 million per year, and since 1993 has been about U.S.$4.4 million. About 90% of the funding comes from MITI, and about one-half goes solely for highly sophisticated research facilities for plant experiments.

Local Government Environmental Research Institute Acid Deposition Projects

The Association of Local Government Environmental Research Institutes (*Zenkoku Kōgaiken Kyōgikai*), the nationwide association of prefectural and metropolitan environmental research institutes, created an Acid Precipitation Survey Section in 1990, and in 1991 began its first acid deposition project, a three-year (1991–1994) nationwide survey of precipitation chemistry.[5] In 1995 another three-year project was initiated, this time focusing on the relation between ambient air pollutants and precipitation pollutants, and on dry deposition. In addition to these two multiyear coordinated programs, local government environmental research institutes engage in self-initiated research projects and research done in conjunction with national government projects such as GERF and the Acid Deposition Survey.

LRT Problem-Framework

No single document neatly expresses Japan's acid deposition problem-framework in its entirety. Instead it is expressed in bits and pieces in a wide array of technical documents. To further complicate matters, the knowledge base of the framework is rapidly expanding and changing. This section offers a snapshot of Japan's LRT knowledge base as it stood in the late 1990s in the four areas of monitoring, emission inventories, atmospheric modeling, and ecological and nonecological impacts. In each subsection, primary research is profiled and the knowledge base derived from these programs presented. The knowledge base, together with methods used to create it, constitute Japan's acid deposition-related LRT problem-framework. In addition, bridging objects and policy relevant implications of what is (or is not) known are discussed.

Monitoring

Monitoring is essentially where the acid deposition problem begins and ends. Precipitation chemistry monitoring in Europe, North America, and Japan revealed the existence of regional-scale, transboundary acid deposition problems. And precipitation chemistry monitoring will reveal their disappearance if and when that dream comes to pass. In this appendix, "monitoring" refers primarily to precipitation chemistry monitoring, i.e., the systematic sampling and chemical analysis of acidic substances in precipitation. Ideally, for comprehensive acid deposition research, simultaneous monitoring of wet deposition, dry deposition, and ambient air quality is desirable. This is seldom realized in practice because of cost, staffing, or other considerations. Wet deposition, or precipitation chemistry, monitoring is most common.

Monitoring programs are essential to both the science and politics of acid deposition. From a scientific point of view, monitoring data provides science-to-science bridging objects such as information on the spatial and temporal distribution of acidic species, precise measurements for validation of long-range transport computer models, and input data for quantifying ecological impacts. From a political point of view, monitoring data provides science-to-policy bridging objects such as confirming the existence of a problem, determining if it is transboundary, and judging the effectiveness of emission control measures. Current monitoring activities in Japan are listed in table A.2. Each is described briefly.

Environment Agency's Acid Deposition Monitoring Network

The EA's National Acid Deposition Monitoring Network (NADMN) is the "official" monitoring network in Japan. As discussed in chapter 8, it was established in 1983. At the end of phase 2 of the Acid Deposition Survey there were twenty-nine sites. At present there are forty-eight sampling stations, including sites on six remote islands (Rishiri, Sado, Oki, Tsushima, Amamiōshima, and Ogasawara).

CRIEPI Monitoring Network

CRIEPI established a monitoring network to investigate wet and dry acid deposition in 1987. During its first phase (1987–1990), CRIEPI operated a twenty-station monitoring network consisting of two types of stations: fifteen stations located in each of fifteen representative climatic regions in Japan, and five stations located on remote islands around Japan (Fujita et al. 1994). A second phase of monitoring was conducted from 1990 to 1996. In the second phase the size of the network was significantly reduced from twenty to seven stations, and the monitoring was

Table A.2
Major monitoring programs in Japan

Monitoring Program	Year Started	Funding Institution	Number of Stations	Comments
(1) NADMN	1983	Environment Agency	48	• First nationwide precipitation chemistry network
(2) CRIEPI network	1987	CRIEPI	20	• Nationwide network • Bulk sampling only
(3) Intensive ground-based monitoring	1990	Environment Agency	8	• Intensive monitoring for long-range transport of pollutants, and acid fog formation
(4) Aircraft and Ship Measurements	1991	Environment Agency/MITI	—	• Special measurements aimed at transboundary movement of pollutants over open oceans
(5) BAPMoN	1976	Meteorology Agency	2	• Part of WMO's GAW program
(6) Local Government Environmental Research Institutes Network	1991	Local Governments	≅ 150	• Nationwide network • Bulk sampling only • Most geographically detailed of all monitoring programs

coordinated with organizations in China (three stations), Korea (two stations), and Taiwan (two stations).

Intensive Ground-Based Monitoring

In addition to routine monitoring carried out by NADMN, the EA selected certain sites for intensive ground-based monitoring of atmospheric trace substances related to acid deposition. These locations are designed to provide more detailed data than obtained at NADMN sites. Funding is provided by the EA's GERF. Sites are selected to fulfill one of three objectives: (1) measurement of pollutants transported from mainland Asia, (2) measurement of background pollutants, and (3) analysis of the formation of acid fog in mountainous areas. Long-range transport monitoring stations were established on the islands of Okinawa (East China Sea), Tsushima (Tsushima Straits between Japan and Korea), and Oki (Sea of Japan). In 1991 additional sites were established at Iriomote and Amami islands in the East China Sea. The Oki Island site was selected to monitor long-range transport from Korea and northern China; the Tsushima Island site to monitor pollutants from Korea and central/northern China; and the Amami, Okinawa, and Iriomote island sites (all part of the Ryūkyū Archipelago) to monitor pollutants from southern China. Background monitoring stations were established on 1,320-meter Mt. Hakkoda in Aomori Prefecture in northern Japan and 1,840 meter Mt. Happo in Nagano Prefecture near the Sea of Japan coast in central Japan. An acid fog monitoring station was established on Mt. Akagi on the western edge of the Kantō Plain in Gunma Prefecture.

Aircraft and Ship Measurements

Aircraft and ship measurements, conducted in Japan with increasing frequency since 1991, supplement and extend data gathered by the various ground-based monitoring networks. MITI's National Institute for Resources and Environment (NIRE) organized the first aircraft measurements (which confirmed long-range transport of air pollutants from continental Asia, and which were described in chapter 9) in 1991 under the direction of Kaneyasu Naoki.[6] On his initial flights northwest of Kyūshū in the Korean Straits, Kaneyasu found very high sulfate aerosol levels below the 1,000-meter cloud base height ($>10\,\mu g/m^3$), far higher than he expected. The high sulfate levels were correlated precisely with a northwest wind off the continent. Repeating measurements in the same location in 1992, and this time coordinating them with shipboard measurements, Kaneyasu and his colleagues again found high sulfate concentrations (>10 μg/m^3) and high sulfur dioxide concentrations (≅ 5 ppb) again correlated with a northwest wind. They also found that the air mass containing these high concentrations measured some 500 kilometers in breadth. The NIRE team announced their data at the Japan Meteorological Society's annual meeting in October 1992. The mass media jumped on the news, and subsequent publicity caused quite a stir, as mentioned in chapter 9. Measurements made thereafter continued to confirm and refine what was discovered the first years, i.e., that plumes with high SO_4^{2-} and SO_2 concentrations sweep off the continent during the winter monsoon season.

Another program similar to NIRE's, the PEACAMPOT (Perturbation by the East Asian Continental Air Mass to the Pacific Oceanic Troposphere) program, also started in 1991 (Hatakeyama et al. 1995a, 1995b). PEACAMPOT is a subprogram of the International Global Atmospheric Chemistry Program (IGAC) within the International Geosphere Biosphere Programme (IGBP). It is more extensive than NIRE's program and consists of aircraft measurements of pollutants over the East China Sea, Yellow Sea, and Sea of Japan, along with simultaneous ground-based observations at various locations. The objectives of the program are to map the four-dimensional spatial and temporal distribution of various trace gases in the troposphere to quantitatively evaluate long-range transport of the pollutants, investigate chemical transformation processes in situ, and provide baseline input data for atmospheric computer models. The program is funded by the EA's GERF and is moving toward establishing ongoing cooperative research with China, South Korea, and Taiwan. All PEACAMPOT missions since 1991 clearly reveal evidence of long-range transport of sulfur and nitrogen compounds from mainland Asia.

Meteorology Agency's BAPMoN Stations

The World Meteorological Organization (WMO) established the Global Atmosphere Watch (GAW) program in 1989 to monitor the changing global atmosphere. It consists primarily of two previously established observation networks: (1) the Global Ozone Observing System (GO3OS),[7] and (2) the Background Air Pollution Monitoring Network (BAPMoN).[8] BAPMoN was established in 1970 and by the end of the 1990s consisted of over 350 stations in over 100 countries. There are two BAPMoN stations in Japan. They are both operated by the Japan Meteorology Agency, and are located at Ryori in Iwate Prefecture in northeastern Japan (established in 1976), and on Minamitori-shima, a tiny island located at the southeastern tip of the Japanese archipelago in the Pacific Ocean (established in 1993). The data from Ryori is the longest and most complete meteorological data set in Asia. The site registered annual average pH levels of 5.2 in the late 1970s. This dropped to 4.7 by 1995.

Local Environmental Research Institutes Acid Precipitation Survey
The Association of Local Government Environmental Research Institutes began a three-year nationwide acid deposition survey in 1991 using filtered bulk samplers. Data was collected at some 150 locations throughout Japan.[9] The Association's survey is by far the most geographically detailed monitoring of precipitation chemistry conducted in Japan to date. It supplements the EA's NADMN and contributes to an understanding of local precipitation chemistry patterns.

Japan's monitoring-related knowledge base (within its larger acid deposition-related problem-framework) developed rapidly in the late 1980s and early 1990s, and by the late 1990s stabilized into a coherent body of "monitoring knowledge." The methodologies used to generate the knowledge consist primarily of the set of sampling and analytic equipment and techniques that underlie precipitation chemistry measurements, ambient air concentration determinations, and dry deposition calculations (see Tamaki et al. 2000). Such equipment and techniques include wet-only and bulk sampling instruments and procedures, ion chromatography and atomic absorption spectroscopy devices and analytic techniques, seasalt contribution estimation formulas, quality assurance/quality control methods, and so on.

Because of Japan's world-renowned technological prowess, it almost goes without saying that as far as the technological aspects of these methodologies are concerned, Japan possesses some of the finest monitoring equipment in the world. Couple this with high work integrity and sophisticated quality assurance/quality control methods, and the result is precipitation chemistry (and related ambient air concentration) data sets that rank among the best in the world. Japan's monitoring knowledge base is sound, reliable, and accurate, and can be treated by policymakers and others with a high degree of confidence.

Although the purposes and methodologies of the various monitoring networks described previously differ, their combined results yield a fairly coherent and consistent monitoring-based picture of the acid deposition situation in and around Japan. The main elements of this picture are as follows:

1. The volume-weighted annual mean pH of precipitation in Japan is about 4.8. (Acid Deposition Survey phase 2 and 3—4.8; CRIEPI Network—4.8; Local Government Network—4.7.)

2. The range of annual average pHs at the various stations is roughly 4.5–5.8 (Acid Deposition Survey phase 2—4.5 to 5.8; phase 3—4.4 to 5.9; CRIEPI Network—4.6 to 5.2.)

3. Precipitation pH tends to be lower in western Japan and higher in eastern Japan. Because annual precipitation is larger in western Japan, the wet deposition of H^+ is therefore significantly larger.

4. CRIEPI estimated the average nationwide wet deposition of nss-SO_4^{2-} to be about 2.5 g/m^2/yr (or 0.8 gS/m^2/yr) in the early 1990s, which was roughly equivalent to the 2–3 g/m^2/yr reported for northwestern Europe for the same time period.

5. There is as yet no strong evidence of increasing acidification of precipitation. For instance, the results of phases 2 and 3 of the Acid Deposition Survey were roughly the same as those found in phase 1.

6. The principal contributor to precipitation acidity is sulfate, as indicated by the S/N (nonseasalt sulfate to nitrate) ratio. The annual average was on the order of 2.5 in the early 1990s. Thus, over the whole of Japan, sulfate (and hence sulfur dioxide emissions) contributed more to precipitation acidity than nitrate (and hence nitrogen oxide emissions). However, between the end of phases 2 and 3 of the Acid Deposition Survey the ratio changed from 2.27 to 1.89, indicating the increasing impact of nitrogen oxide emissions.

7. The S/N range for the Acid Deposition Survey phase 2 was 1.4–5.0, and for phase 3 was 1.02—4.35. The Sea of Japan coast region had a higher average ratio than the Pacific coast region during the winter. Thus, during the winter nss-SO_4^{2-} plays a greater role in the acidity of precipitation on the Sea of Japan coast than on the Pacific coast.

8. Wet and dry deposition of sulfur appear to be roughly equal.

9. Volcanoes play a major role in the acidification of precipitation in Japan.

10. Long-range transport of acidic pollutants from mainland Asia is established beyond question. Most pollutants are transported during the winter. However, the amounts and specific sources remain in question.

Japan's expert monitoring knowledge as it stood circa 2000 yielded the following high-quality science-to-policy bridging objects (with the bridging object's type listed in parentheses): (1) precipitation in Japan is acidic—at about the same levels as Europe and North America (extent-and-intensity), but there was no sign of increasing acidity (trends); (2) precipitation is more acidic in western Japan than eastern Japan (distribution); (3) sulfate contributes more to acidity than nitrate, but in large urban areas the contribution of nitrate is relatively higher (character); (4) volcanoes contribute significantly to Japan's acidic precipitation (character); and (5) long-range transport of acidic pollutants from mainland Asia is occurring (existence). These objects are based on solid data and are a solid platform from which political actors can determine their interests and make utility calculations.

Emission Inventories

East Asia-wide emission inventories, not Japanese domestic inventories, are the focus of this section.[10] Three separate emission inventories for East Asia appeared within the span of only one year in 1991. The first was compiled by the World Bank's RAINS-Asia project.[11] Japanese organizations compiled the other two, as briefly explained next.

CRIEPI developed the first SO_2 and NO_x emissions inventory to focus exclusively on East Asia. The SO_2 inventory (Fujita et al. 1991), and the NO_x inventory (CRIEPI 1992) were compiled using energy consumption and industrial activity data. Emissions from fossil fuel combustion were estimated on an 80 kilometer x 80 kilometer grid basis (the approximate size of a 1° latitude x 1° longitude grid in the center of the East Asian region). The result was a gridded emissions map for East Asia (Japan, China, Taiwan, South Korea, and North Korea). After completion of the two inventories, CRIEPI has not actively continued emissions inventory work. However, the inventories are periodically updated.

CRIEPI also created the most comprehensive estimate of volcanic emissions to date (Fujita 1993b). Spectroscopic data from 1976–1989 was combined with an empirical volcano classification system to estimate the annual emission rates of SO_2 for eruptive and noneruptive volcanoes. In 1988 the estimated annual SO_2 emissions from twelve active volcanoes was 0.5 to 0.6 TgS per year. This accounted for approximately 7% of global volcanic SO_2 emissions. Since the estimate for anthropogenic emissions in Japan was about 0.5 TgS per year in 1986, this implied that, averaged over the whole of Japan, volcanic and anthropogenic emissions were roughly equivalent. Of the total volcanic SO_2 emissions, about 80%, or 0.4 TgS, emanated from Kyūshū island, with by far the largest contributor being Sakurajima volcano near the southern tip of the island. Sakurajima injects about five times more SO_2 into the atmosphere than local emission sources. Obtaining more accurate estimates of volcanic emissions from Sakurajima and other volcanoes is currently an active area of research.

The most comprehensive Asia-wide emissions inventory compiled by Japanese researchers was carried out by the National Institute of Science and Technology Policy (NISTEP) of the Science and Technology Agency. Results were first published in Kato et al. (1991), and summarized in Kato and Akimoto (1992). The NISTEP group determined anthropogenic emission inventories for SO_2 and NO_x for twenty-five countries in Asia east of Afghanistan for the years 1975, 1980, and 1987 based on fuel consumption, sulfur content in fuels, and emission factors for fuels in various emission categories. Detailed subnational (province and region) calculations were made for the two largest countries, China and India. The total SO_2 and NO_x emissions for the entire Asian region in 1987 were 14.5 TgS and 4.7 TgN, respectively. The SO_2 and NO_x emissions for East Asia (China, Japan, South Korea, North Korea, and Taiwan) in 1987 were 11.7 TgS and 3.3 TgN, respectively. In 1994 a 1°

$\times$ 1° resolution SO_2, NO_x, and CO_2 inventory based on the NISTEP inventory was published (Akimoto and Narita 1994).

Japan's East Asian emissions-related knowledge base (within its larger acid deposition-related problem-framework) is still weak. To date it relies on fuel consumption data along with certain fuel-specific and technology-specific parameters. Rough socioeconomic factors are used to distribute emissions spatially and temporally. There is as yet little use of actual measurements of emissions to compile or verify inventories. In addition, detailed information on location of point sources, stack heights of point sources, diurnal variations of emissions, and so on, is generally lacking. Although the SO_2 and NO_x inventories compiled to date generally agree with one another, which gives them some credibility, their overall quality is still coarse and crude because the methodologies are still coarse and crude. Therefore, Japan's emissions knowledge base can be treated by policymakers with a measure of confidence as to overall magnitudes and patterns of emissions, but not as to details. The quality of "emissions knowledge" falters not for lack of effort but for lack of precise statistics in the East Asian region as a whole. Japan is limited in its ability to collect emissions information in other countries. Gathering the data necessary to compile regional inventories can be intrusive and politically sensitive.

The main elements of the "East Asian emissions picture" created by Japanese researchers is as follows:

1. Total emissions from East Asia rival those of Europe and North America. Total East Asian emissions of SO_2 in 1990 were approximately 11–12 TgS per year (= 22–24 $\times 10^6$ tonnes SO_2).

2. Emissions of acidic substances from China are huge, and generally increasing. All inventories agree on this central point. As of 1990, estimates of emissions of SO_2 in China were about 9–10 TgS per year (= 18–20 $\times 10^6$ tonnes SO_2), or about 90% of total SO_2 emissions in East Asia.

3. Tokyo is East Asia's NO_x hot spot. It is the single largest NO_x source region in Asia.

4. Volcanic emissions are extremely important but still highly uncertain. The only estimate to date on total volcanic emissions in Japan places them roughly equivalent to domestic anthropogenic emissions.

5. Other nonanthropogenic emissions (such as sulfate from marine sources) have not yet been estimated for the East Asian region. Also, inventories for acidic species other than SO_2 and NO_x are still uncertain, though work is underway on inventories for ammonia, VOCs, and others.

6. Much work remains on making existing inventories more accurate and precise. Efforts are underway to include information such as location of point sources, heights of sources, diurnal variations of emissions, etc.

Japan's expert East Asian emission knowledge as it stood circa 2000 yielded the following 'reliable' science-to-policy bridging objects (the bridging object's type is listed in parentheses): (1) China's emissions are huge and increasing (trends), (2) East Asian emissions are large due to the phenomenal industrial growth in the region and are expected to continue to climb in the long-run (trends), and (3) nitrogen oxides emissions seem to be on the increase due to increasing motor vehicle use (trends). As we saw in chapter 9, trends bridging objects such as the above are drivers of Japan's policymaking on transboundary air pollution in East Asia.

There is common agreement that the next step in emissions inventory work involves cooperatively developing national inventories based on similar methodologies that can be combined into regional inventories (ESCAP 1999). Japanese policymakers' long-term goal is to bring emissions inventory work under EANET's umbrella.

Long-Range Atmospheric Transport Models

There is a wide range of research in Japan on atmospheric transport of acidic pollutants. This section focuses solely on long-range transport (LRT) computer models. Shorter range models that analyze Japan's domestic, small-scale air pollution and acid deposition problems are not discussed; nor is experimental work on acidic pollutants in the atmosphere. Active LRT modeling projects in Japan are listed in table A.3. Three representative models are discussed here (CRIEPI, MRI, and Uno).

CRIEPI Model

CRIEPI became interested in local-scale air pollution computer models in the 1960s, and began developing an LRT model in 1987. This work continues today (Ichikawa and Fujita 1993; Ichikawa and Fujita 1994; Ichikawa et al. 1994; Ichikawa and Fujita 1995; Kohno 1997). The most significant application of CRIEPI's model to date is estimation of the contribution of East Asian countries to wet deposition of nonseasalt sulfate in Japan. The contributions to Japan as a whole and to four climate-based regions (Sea of Japan coast, Pacific coast, Inland Sea/central mountain district, and northern Hokkaidō) of emissions from Japan, China, North and South Korea, and Taiwan were computed. The breakdown in SO_2 emissions of the five countries in the simulation was as follows (in 10^6 tonnes): China (18, or 85% of total emissions), the Koreas (1.9, or 10%), Japan (0.9, or 4%), and Taiwan (0.3, or 1%); for a total of 21×10^6 tonnes. Volcanic emissions (only for the main islands of Japan) were equivalent to about 1.5×10^6 tonnes, or about 7% of East Asian anthropogenic emissions. CRIEPI applied the model to estimating the contribution of East Asian countries to deposition in Japan. Besides the full-year calculation, the year was divided into two periods—summer (October through March) and winter (April through September)—and the seasonal differences investigated. The total contribution of the four country-regions (Japan, China, North and South Korea, and Taiwan) was set at 100%.

Model results yielded the following conclusions:

1. China contributed about 50%, Japan 33%, the Koreas 17%, and Taiwan almost nothing to total annual nss-SO_4^{2-} wet deposition in Japan during the one-year period from October 1988 through September 1989.
2. Of total annual wet sulfate deposition, China contributed about 60% to Sea of Japan coast deposition, about 50% to Pacific coast deposition, and about 35% to Inland Sea/central mountain district deposition.
3. China contributed more in winter than summer; Japan contributed more in summer than winter; and there was no seasonal tendency in the contribution from the Koreas or Taiwan.
4. Comparing continental (China and the Koreas) contributions, they were higher to the Sea of Japan coast than to the Pacific coast. The contribution was especially high in winter—over 80%. The contribution of Japanese domestic emissions was highest in the Inland Sea/central mountain district—over 50%.

Table A.3
Long-range transport computer models in Japan

Name	Year Initiated	Main Funding Source	Model Type(s)	Comments
CRIEPI	1987	CRIEPI/MITI	Lagrangian	• Currently developing a "hybrid" combined Lagrangian (large-scale)/Eulerian (short-scale) model
MRI	1990	Environment Agency (GERF)	Lagrangian	• Transport model coupled with operational weather forecasting model
Uno	1993	Environment Agency (GERF)	3-D Eulerian	• Adaptation of STEM-II model • First 3-D model to use complex atmospheric chemistry in Japan • Originally developed at NIES by Uno Itsushi, who is now at Kyushu University
Katatani	1986–1988	IEMAJ & Ministry of Education	Lagrangian & 3-D Eulerian	• Originally developed by the Industrial Environmental Management Association of Japan, now independently developed by Katatani Noritaka of Yamanashi University
Kitada	1990	Ministry of Education	3-D Eulerian & Lagrangian	• Developed by Kitada Toshihiro of Toyohashi University of Technology

Table A.3
(continued)

Name	Year Initiated	Main Funding Source	Model Type(s)	Comments
				• Models not now under active development in relation to acid deposition
Okita	1987	Ministry of Education	3-D Eulerian	• Developed by Okita Toshiichi of Obirin University • One of first models to demonstrate transboundary transport of pollutants

Note: Perhaps typical of the Japanese avoidance of self-assertion, none of the Japanese models in this table have individual names or acronyms. Thus, all acronyms are the author's creation to facilitate comparison.

MRI Model

The Meteorological Research Institute (MRI) of the Meteorology Agency initiated development of an LRT model in 1990. The model is a combination of a highly sophisticated meteorological submodel and an "ordinary" Lagrangian submodel, and is designed to estimate both dry and wet deposition of SO_2 and SO_4^{2-} in the East Asia region (Applied Meteorology Research Department 1995; Sato et al. 1999). The MRI model's strength lies in the fact that it is an adaptation of a full-scale operational weather forecasting model. The meteorological submodel is used to predict meteorological field variables such as wind speed and direction, pressure, temperature, humidity, etc. for incorporation into the Lagrangian submodel.

Numerical calculation of such meteorological variables is particularly critical in East Asia because of the large area occupied by ocean surface where only sparse observational data are available. The first model application to East Asia used the NISTEP emissions inventory. Japanese emissions were excluded to simulate the impact of continental long-range transport on acidic deposition in Japan. A full-year (with a ten-minute time step) using 1985 emissions and meteorological data was simulated. The model calculated transboundary dry and wet deposition in Japan, and compared these values to the EA's monitoring network wet deposition values. The following were the key conclusions:

1. Simulated annual dry deposition tended to decrease from west to east, and south to north. Simulated annual wet deposition showed no such tendency. The highest simulated total (dry + wet) depositions were in Kyūshū in the southwest, and the lowest were in the Tokyo area. Depositions were moderately high on the Sea of Japan coast and in Hokkaidō.
2. Simulated SO_4^{2-} wet deposition was compared to monitored SO_4^{2-} wet deposition at four representative urban receptor grids (Osaka, Nagoya, Tokyo, and Tomakomae in Hokkaidō), which overlapped with the EA's paired urban and rural monitoring stations. The simulated values averaged around 0.05 gS/m^2/yr, whereas the monitored values ranged between 0.6–1.3 gS/m^2/yr. This implied that continental anthropogenic emissions contributed on the order of only 10% or less to Japan's total annual SO_4^{2-} wet deposition.

Uno Itsushi's Models

The EA's National Institute for Environmental Studies (NIES) has long been involved in air pollution modeling activities, especially the development of urban air pollution models and medium-range transport models. It wasn't until 1993 under the direction of Uno Itsushi of the Atmospheric Environment Division that work began on a long-range transport model when he adapted the Eulerian STEM-II model to East Asia. Uno moved to Kyushu University in 1998 where he continues his work. He has also developed a long-range transport model coupled with the Regional Atmospheric Modeling System (RAMS) originally created at Colorado State University. Publications include: (Uno 1995, 1997; Uno et al. 1998a, 1998b, 1998c; Uno and Sugata 1998).

In its first application, Uno's STEM model successfully simulated elevated sulfate and nitrate concentrations in high resolution aerosol monitoring data from June 1991 to February 1992 at Tsushima Island. The model demonstrated that the location of the *baiu* rainy season meso-front played a significant role in transport of pollutants from Asia to Japan in a June 1991 event, and that the passage of a low pressure system from the East China Sea to the Pacific Ocean resulted in a February 1992 event. The RAMS model, which utilizes real-time meteorological data, also successfully simulated the June 1991 event. Uno's models have since simulated sulfate and nitrate aerosol data at other locations, and reveal the existence of strong sulfate and nitrate pollutant gradients on the north side of the *baiu* front. The front typically lies in a line between southern China and southern Japan with South Korea and Kyūshū generally on the north side of the front. This implies that the location of the *baiu* front is a prime determinant of sulfate loading in South Korea and southern Japan during the rainy season. The front acts as a trap for air pollutants emitted in China and on the Korean Peninsula. Uno's models, more than any others in Japan, capture the complex physical and chemical processes involved in long-range transport of pollutants in East Asia.

Japan's long-range transport computer modeling-related knowledge base (within its acid deposition-related problem-framework) is developing with stunning speed. The methodologies that underlie "LRT computer simulation-based knowledge" include meteorological data analysis techniques, vertical diffusivity profile formulas, turbulent diffusion models, atmospheric chemistry schemes, physical and chemical descriptions of precipitation processes, calculation of deposition velocities, etc. In addition, there are computational methods used to perform the calculations demanded by these models. The methodologies in Japan are generally not as advanced as in the West; however, Japan is rapidly approaching parity. The quality of "LRT modeling knowledge" is good, but since it is still developing, policymakers cannot as yet place a high degree of confidence in model results. Even so, Japan has the largest number and highest quality LRT models in Asia.

The first essential policy-related question that LRT models can help Japanese policymakers answer is: "Is transboundary transport of pollutants occurring?" Or more specifically: "According to LRT models, are acidic pollutants emitted in other countries being deposited in Japan?" The answer of all models to both questions is affirmative. Hence, model results reinforce the "import of air pollutants to Japan" existence bridging object.

The next question that policymakers would like LRT models to answer is: "How much of Japan's acid deposition is due to non-Japanese sources?" Here there is no agreement yet among the LRT models, and hence no bridging object with a strong message. The models' answers vary widely. The MRI model estimated that less than 10% of the total annual SO_4^{2-} wet deposition in Japan is due to continental sources. The CRIEPI model estimates completely the opposite—that about 67% of the total annual SO_4^{2-} wet deposition is due to continental sources. The IEMAJ model (not discussed previously) estimates that about 30–40% of Japan's acid deposition (sulfate and nitrate) is due to continental sources.[12]

A third essential policy-related question that LRT models can theoretically answer relates to specific source-receptor relationships: "What non-Japanese sources contribute how much to what receptor points in Japan?" Here again there is no agreement, and hence no bridging objects. Comparing three East Asian models, Streets et al. (1999, 138) state that China's contribution to Japan's total sulfate deposition was 17% in the RAINS-Asia model, 3.5% in a Chinese model, and 50% in the CRIEPI model. Even though total sulfate deposition in the whole of East Asia was similar in these models, they displayed highly disparate estimates for the China/Japan source-receptor relationship. One of the few points of agreement among Japanese and other models is that continental emissions contribute significantly to winter deposition of SO_4^{2-} on the Sea of Japan coast, and that domestic emissions contribute significantly to summer deposition, especially on the Pacific coast. Also, to the degree that models consider volcanic emissions, all show that volcanic emissions are a significant contributor to annual sulfate loading in Japan.

Just as all locations cannot be monitored, and just as all emission sources cannot be tabulated, so too with LRT modeling, all atmospheric processes cannot be modeled. There are numerous "problem areas" in the content of Japan's LRT computer simulation-related knowledge base. These include the coarse quality of emission inventories used by the models; lack of inclusion of emissions from far eastern Russia; uncertain precipitation distribution patterns; inability to accurately characterize precipitation amounts and intensities, uptake rates of sulfur dioxide and sulfate over the open oceans surrounding Japan, cloud formation and in-cloud scavenging processes, and dry deposition velocities; and lack of consideration of the *kōsa* phenomenon (i.e., long-range transport of alkaline dusts). The ultimate validation of LRT models is direct comparison of model-calculated values with actual measured values. However, in Japan observational data sets are still too small, and model-calculated values too rough to make detailed comparisons. In summary, as far as modeling is concerned, the consensus core is small and contested periphery large. Thus, to date there has been considerable difficulty constructing effective bridging objects from computer simulation results.

Ecological and Nonecological Impacts

There is a tremendous range of research in Japan attempting to assess types and locations of acid deposition impact. It is beyond the scope of this book to describe on all such research. However, the following is touched upon: (1) ecological impacts—inland waters, soils, crops, and forests; and (2) nonecological impacts—cultural heritage and materials.[13] The main impacts-related programs are: the Inland Waters Survey, Soils Survey, and Forests Survey within the Acid Deposition Survey; and the Watershed Chemical Input-Output Study (analyzing two watersheds in detail), the Acid-Neutralizing Capacity of Surface Soils Study (analyzing the ANC of soils throughout Japan), and the Effects of Acidic Deposition on Forests Program (concentrating on the decline of Japanese cedar) within CRIEPI's Acid Deposition Project. The Ministry of Agriculture, Forestry, and Fisheries (MAFF) also established a Nationwide Monitoring of Forest Damage and its Relation to Acid Deposition (*Sanseiu Nado Shinrin Higai Monitaringu Jigyō*), which is patterned after Europe's International Cooperative Program on Assessment and Monitoring of Air Pollution Effects on Forests. In addition, numerous smaller studies conducted under the EA's GERF contribute to impacts research, as does a wide range of research undertaken at universities and at local government environmental research laboratories.

Japan's ecological and nonecological impacts-related knowledge base (within its acid deposition-related problem-framework) is just beginning to mature. Methodologies used to generate "impacts knowledge" includes methodologies related to soil analysis, vegetation surveys, lake water chemistry, and damage to art treasures, to name just a few. One statement that can be made about impacts work up to the late-1990s is that much of it is survey type of research. In other words, the methodologies are generally oriented to gaining an overview of various possible impacts and impacted areas rather than delving into detailed cause-and-effect mechanisms. Some of the main elements of the "impacts picture" in Japan are as follows:

Inland Waters

• According to phase 2 of the Acid Deposition Survey, most water bodies in Japan have pH values of around 7. No annual trends of increasing acidification were identified in phases 2 or 3.

• According to CRIEPI's watershed research, the input-output budget of chemical species of the two watersheds studied suggests that the major source of H^+ was microbial decomposition of organic matter and respiration of plant roots in the soil, not wet deposition.

• There is concern about acid shock from snowpack meltwater. According to the Acid Snow Survey of phase 2 of the Acid Deposition Survey, meltwater in early spring was acidic in the two locations studied, and had a slightly lower pH than rain or snowfall. However, no evidence was found that the acidic meltwater noticeably affected river and streamwater.

• One early GERF project on acid deposition-sensitive watersheds concluded that acidification of the most weakly buffered lakes and rivers in Japan would not occur at all, or would not occur for at least several decades, even if present levels of acid deposition loading continued.

• However, another GERF project on inland water bodies in the Japanese Alps indicated potential acidification (Kurita et al. 1993), implying acidification of inland waters is still controversial.

• To help resolve the controversy, phase 3 of the Acid Deposition Survey used a lake acidification model to study the effects on five lakes. The model indicated that at present levels of acid deposition the possibility of acidification was small; however, if deposition of acids dramatically increased, then damage to the most susceptible lake could occur within thirty years.

• Overall, the current consensus is that Japan's inland waters are not in any immediate danger of acidification.

Aquatic Organisms

• Laboratory research by the National Institute of Fisheries has shown that there are detrimental effects of acid deposition and heavy metal mobilization to freshwater fish.

• However, these types of effects have not been verified in any natural freshwater systems.

• Overall, the current consensus is that Japan's freshwater aquatic organisms are in greater danger from causes other than acidification of inland waters such as land use changes and development projects.

Soils

• According to phases 2 and 3 of the Acid Deposition Survey's soil surveys, there is no clear-cut evidence of major changes in the physical-chemical properties of soils in Japan that can be attributed to acid deposition.

• Also, according to experiments on the effect of artificial acid rain on various types of soil conducted as part of phase 2 of the Acid Deposition Survey, the levels of pH at which clear evidence of soil acidification appears are far below present annual average pH values of rainfall in Japan.

• According to CRIEPI and other research, the classification system used in the Environment Agency's Map of Soil Susceptibility to Acidification seems basically accurate, but individual soils may deviate significantly from the standard.

• Overall, the current consensus is that Japan's soils are not significantly acidified from anthropogenic sources, nor is there a detectable trend of acidification.

Crops

• Little in the way of adverse effects to crops or the agricultural environment was found in either indoor or outdoor research conducted under the GERF and CRIEPI programs.

• Overall, the current consensus is that Japan's crops are not affected by current levels of acid deposition.

Forests

• According to phases 2 and 3 of the Acid Deposition Survey's forest surveys, there is evidence of tree decline in some locations. However, in most cases the causal factors can be identified and are not related to acid deposition. In a few locations, however, the causes cannot be identified. Thus, acidic precipitation remains a possible cause of the forest decline.

• According to CRIEPI surveys of damage to Japanese cedars, the presence of high oxidant pollutant levels may be the dominant cause of damage, and low precipitation levels may be an exacerbating factor. This casts doubt on acid deposition as a major factor.

• According to CRIEPI experiments on acid rain's effect on tree seedlings, acidified precipitation at levels currently found in Japan (with minimum pH values around 4.0) do not induce any direct or significant impact on the tree seedlings species tested.

• A MAFF nationwide Forest Decline Survey has to date found no clear link between forest decline and acid deposition. Other more site-specific surveys, such as at Nikkō National Park, echo this conclusion.

• However, there remain locations of forest decline throughout Japan where acid deposition cannot be ruled out as at least a contributing factor.

• Overall, the current consensus is that there is no solid consensus on the impact of acid deposition on Japan's forest. Forest decline and damage remain the most controversial area of acid deposition impact in Japan.

Cultural Properties and Materials

• Research in this area is still young and no strong conclusions can yet be made, but to date there do not seem to be significant impacts to cultural properties or materials due to acid deposition alone.

Japan's expert "ecological and nonecological impacts knowledge" as it stood circa 2000 is controversial, and its quality modestly adequate. It has, however, yielded the following important science-to-policy impacts and risk assessment bridging objects: (1) there is no unequivocal evidence of large-scale ecological or nonecological impacts in Japan due to acid deposition, (2) impacts to crops have all but been ruled out, and (3) impacts to inland water bodies and aquatic organisms is not a present worry but could become serious in the future. The current focus of impacts research centers on soils, forests, and cultural properties and materials. There are, however, dissenting voices among researchers on the certainty of some of these conclusions. Some are convinced that Japan is immune to large-scale ecological damage. Others say Japan is susceptible to small-scale impacts, but not large-scale. Still others argue that it is too early to make any definitive conclusions. Thus, policymakers are generally taking a cautious approach to the degree of trust they place in the present impacts-related knowledge.

Notes

Chapter 1

1. A few examples of Japanese "rain words," which are most often very difficult to translate because of their nuanced meaning, include the following: *konuka ame* (fine rice bran rain; a very light rain that just barely gets you wet), *samidare* (fifth-month rain; a type of drizzle that occurs during the late spring rainy season), *shigure* (momentary rain; late fall or early winter burst of rain that occurs in a small area and over a short period of time, usually associated with being in the mountains); *gōu* (powerful rain; a downpour, often destructive, in the late summer or early fall typhoon season); *jiu* (luxuriant or beneficial rain; a rain from the gods such as after a long drought); *namida ame* (tears rain; a light rain that reflects one's mood more than the meteorological phenomenon); and *yarazu no ame* (prevented from going rain; a phrase used by women to refer to a rain that prevents a lover, who wants to leave but who the woman wants to be with, from departing).

2. The Limited Test Ban Treaty of 1963 was the first international convention related to transboundary air pollution. It totally banned atmospheric testing of nuclear weapons, and partially banned certain forms of underground testing. The discovery of global deposition of strontium-90 was a major factor leading to the ban on atmospheric testing. (For more details, see Soroos 1997, 83–109.)

3. For a comparative analysis of Europe, North America, and East Asia, see Wilkening (2004).

4. For background on the Edo period, see Nakane and Oishi (1991).

5. If we add to our consideration *resource-related* sustainability crises, we could perhaps add the period leading up to World War II as another sustainability crisis in Japan's modern history. Japan's lack of, or perceived lack of, access to resources vital to industry, especially energy resources, was one factor driving the aggression that resulted in the Pacific War. Today, attaining "resource security" is a fundamental tenet of Japan's foreign policy. Given Japan's overwhelming dependence on imported raw materials and foodstuffs, the country is in an almost perpetual resource-related sustainability "crisis." In this book,

however, only sustainability crises related to industrial pollution are considered. (I thank an anonymous reviewer for reminding me of the prewar resource-related sustainability crisis.)

Chapter 2

1. The WCED defined sustainable development, not sustainability. The two terms are similar but not necessarily synonymous. WCED's 1987 report was highly influential and popularized the term "sustainable development." I prefer "sustainability." For critiques of the sustainable development terminology, see Lohmann (1990), Lélé (1991), and Nagpal (1995).

2. No attempt is made in this book to review the science studies literature. Such reviews can be found in Miller and Edwards (2001, 11–15), Jasanoff et al. (1996), and Fujimura (1996, 237–243).

3. An environmental problem-framework (e.g., an acid deposition problem-framework) is different from an environmental science (e.g., acid deposition science). While the two are similar, a science is the larger entity. Frameworks are subsumed within it. A science houses the full historical collection of scientific information relevant to an environmental topic. Not all of this information is necessarily relevant to constructing a problem-framework. In addition, a desire to understand the underlying dynamics of natural phenomena generally motivates a science; whereas addressing a specific environmental problem motivates a problem-framework. A science also implicitly contains the institutional and socialization mechanisms that govern the practice of the science. Although the two terms are close in meaning and sometimes interchangeable, I maintain a distinction between a *science* and a *problem-framework* to emphasize the general and ostensibly apolitical nature of the former and the more specific and generally political nature of the later.

4. I use the word *knowledge-world* deliberately in reference to Karl Popper's "world 3" (Miller 1985). Popper, a philosopher of science, argued that there are three worlds—a world of physical objects or states (world 1), a world of consciousness or mental states (world 2), and a world of the objective contents of thought (world 3). World 3 contains true theories and false ones, open problems and closed ones, the content and state of discussions and critical arguments, conjectures and refutations, and, the contents of journals, books, and libraries. Popper defines "world 3" with the purpose of trying to prove that there is such a thing as "objective knowledge." Although his arguments have been roundly criticized, I adopted his notion of a mixed bag of knowledge that reflects the *state* of knowledge at any given time more than the truth of the knowledge. Popper's failed attempt to establish the existence of objective knowledge on the basis of his "world 3" serves well to illustrate the fact that what characterizes the "knowledge-world" of a problem-framework is not its objectiveness or truthfulness per se, but its attempt at objectiveness. It is pursuit of objective truth that empowers a problem-framework in the political world.

5. Social worlds are defined as "loosely or rigidly structured units in which people share resources and information . . . characterized by a commitment to common assumptions about what is important, and what should be done" (Garrety 1997, 730). It is a term that addresses the collective social negotiation of meanings.

6. I am not arguing that there is an absolute demarcation line between scientists and the support apparatus, or between scientists and policymakers. Besides the work of Fujimura, and Star and Griesemer, many others argue forcefully that science is co-constructed; see, for instance, Van Eijndhoven and Groenewegen's (1991) analysis of how "flexible argumentation chains" link scientific data, expert interpretation, and policy meaning. See also Miller and Edwards (2001) and Jasanoff (1987, 1990).

7. The epistemic community term was apparently first coined by Burkart Holzner (1972). John Ruggie (1975) introduced it into international relations literature, where it was used by Ernst B. Haas (1990) and Peter M. Haas (1989, 1990a, 1990b, 1992a, 1992b, 1992c, 1993a, 1993b, 1997). Peter Haas defines an epistemic community as "a network of professionals with recognized expertise and competence in a particular domain and an authoritative claim to policy-relevant knowledge within that domain or issue-area" (Haas 1992b, 3). He identifies four constituent elements of epistemic communities: (1) shared normative and principled beliefs (i.e., a set of beliefs that form a value-based rationale for social action by its members), (2) shared causal beliefs (i.e., belief in the validity of certain specific cause-and-effect relationships), (3) shared notions of validity (i.e., acceptance of certain truth-testing methods such as the scientific method), and (4) a common policy enterprise (i.e., a common focus on a particular problem area). This definition has attracted considerable attention, particularly from those researching environmental and scientific issues in the international arena, but it has also attracted considerable confusion. Many international relations scholars severely criticize the epistemic community concept as it currently stands. In my modified definition, constituent elements 1 and 2, the source of most of the controversy, are dropped and elements 3 and 4 are emphasized.

8. Some readers might wonder why I do not use terms such as "scientific framework" and "scientific community" since I seem to be dealing primarily with scientists and scientific knowledge. Besides the fact that these terms already enjoy widespread use in contexts very different from the one addressed in this book, scientists are not the only possible constituents of an expert community and scientific knowledge is not the only possible constituent of a problem-framework. Expert communities and problem-frameworks house experts and expert knowledge that may extend beyond the natural sciences. When addressing multidisciplinary environmental problems, expert communities can include, for instance, professionals from law, psychology, sociology, anthropology, economics, political science, engineering, and other disciplines. Problem frameworks can include expert knowledge derived from these various disciplines. In the case of Japan's acid deposition history up until recently, expert communities were almost

exclusively composed of natural scientists and problem-frameworks almost exclusively composed of scientific knowledge derived from the natural sciences. But this is beginning to change. I have opted to use a terminology that I believe is inclusive beyond the natural sciences.

9. This definition is taken from David Easton (1953, 129). It, of course, parallels the classical definition of microeconomics as the efficient allocation of scarce resources.

10. For example, Van der Sluijs et al. (1998) call "climate sensitivity" (response of climate models to a doubling of atmospheric CO_2) in the global climate change field a boundary object (bridging object in my terminology). In fact, they argue that climate sensitivity is a special class of boundary objects, called "anchoring devices," that fixes the scientific basis for policy debate and makes complex science more transparent to policymakers.

11. My policy-activist grouping corresponds exactly to what Peter Haas (1992b) defines as an epistemic community. In a critique of Haas's epistemic community concept, James Sebenius (1992) very succinctly defines an epistemic community as a "de facto natural coalition" whose task is to build a winning coalition of support behind its preferred policy choice. This corresponds to how I define policy-activists. This group is a de facto natural coalition, first because of its privileged position in relation to the core of sophisticated technical knowledge related to the problem area, and second because of its value-based goal. Thus, Haas's epistemic community becomes a subgroup within my broader definition of expert community.

Chapter 3

1. The first scholarly works that attempted to place human-environment interactions (such as acid deposition) in global historical perspective appeared in the 1950s. One of the earliest was a pathbreaking collection of writings titled *Man's Role in Changing the Face of the Earth* (Thomas 1956). Another, focusing only on the western world, was C. J. Glacken's (1967) massive tome *Traces on the Rhodian Shore*. In the same geography tradition as *Man's Role* and Glacken's work is a more recent international effort by the Earth Transformed Project focusing on the past 300 years, *The Earth as Transformed by Human Action* (Turner 1990). All of these works are environmental or "green" histories. In recent years a wealth of green histories have burst forth; see, for instance, the world histories of Ponting (1993) and Simmons (1996). This literature attempts to provide overarching anthro-environmental contexts in which specific problems such as acid deposition can be placed.

2. See, for instance, Barrie (1986), Whelpdale et al. (1988), Wilkening et al. (2000), and Wilkening (2001).

3. Wherever sensitive ecosystems coincide with industrial development future hot spots may appear. The level of investigation in the rest of the world outside the three hot spot regions is still low, and publications few, but potential trouble

spots include Southeast Asia, South Asia, South Africa, and South America (Rodhe 1989; Rodhe and Herrera 1988).

4. Further information on Japan's relationship to nature can be found in Asquith and Kalland (1997), and Brecher (2000).

5. Readers desiring further information on Japan's geography and natural environment should consult the nine-volume *Kodansha Encyclopedia of Japan* (1983). This is the definitive English language encyclopedia on Japan and is now available on CD-ROM. I consulted the encyclopedia to help construct the portrait of Japan's natural setting given in this section. Other sources include Reischauer (1981, 1–36) and those cited in the text.

6. I thank Satake Kenichi for bringing to my attention Japan's natural richness of acidified environments. Data on the acidity of Japanese lakes in this section comes from Satake and Saijo (1974) and Satake et al. (1995) and is discussed in more detail at later points in the text.

7. Information on Japan's climate derives from Fukui (1977), Kodansha (1983, Vol. 1, 325–328), and Katoh et al. (1990, 45–46). Fukui's book is the most comprehensive work in English on Japan's climate.

8. The yellow color of these soil particles or sands is also the origin of the name of the mighty Yellow River of China and the Yellow Sea where the Yellow River ends its journey. So too, the legendary first emperor of China was the Yellow Emperor.

9. It is believed that this term originated in ancient China from the fact that plums, the first tree to bloom after the winter snows and the first fruit to be harvested, were ripe by the time of the plum rains. Japan imported the word from China and called it *bai-u*. In Japan the plum is prized for its fruit, which, unlike the sweet plum of the West, is inedible even when picked ripe from the tree. It is salted and pickled to become the famous deep red Japanese *umeboshi* (dried plum). The yellow coloration of the original fruit becomes a deep red upon pickling and has a pH of about 2.0, about equivalent to the lowest pHs ever recorded in Japan.

10. An interesting article that echoes many of my conclusions about the differences between Japanese and Western science is Motokawa Tatsuo's essay "Sushi Science and Hamburger Science" (Motokawa 1989).

11. For overviews of Japan's environmental history, see Gresser et al.'s (1981) *Environmental Law in Japan*, which is an outstanding survey of Japan's environmental history from a legal perspective for the period 1868–1980; works by Ui Jun of Okinawa University, one of the best-known analysts, activists, and commentators on Japan's present and past environmental problems (Ui 1991a, 1991b, 1992); a history of air pollution control from the Meiji period to about 1985 (Nishimura 1989); a detailed chronology of pollution in Japan from 1600 to 1975 (Iijima 1979); a history of pollution in Japan by Kamioka Namiko (1987); Margaret McKean's (1981) fine work on the history of citizen environmental groups; Huddle and Reich's (1987) description of Japan's postwar environmental "crisis"; and Jeffrey Broadbent's detailed analysis of local

environmental politics in Oita Prefecture (Broadbent 1998). I consulted these sources. In addition, other literature is referenced in the text.

12. Although this division is based on my personal reading of the environmental history of Japan, it basically conforms to that of other commentators. For instance, Gresser et al. (1981, 3) divide Japan's environmental history into four stages: (1) prewar (1868–1945), (2) postwar, ending with enactment of the Basic Law for Pollution Control (1945–1967), (3) implementation of the Basic Law and the administrative flurry after the four major pollution trials (1967–1975), and (4) the present, which in their book stems from 1975 to about 1980. An NGO pamphlet prepared for the United National Conference on Environment and Development (UNCED) in 1992 divided the postwar environmental history of Japan into four periods: (1) worsening of pollution and birth of citizen movements (1945–1970), (2) antipollution public opinion and progressive environmental policy (1970–1977), (3) regression in policy and new problems (1978–1987), and (4) growing global environmental problems (1988–present) (92 NGO Forum Japan 1992).

13. All of these works are fairly short, and devote only a page or two to the period before the moist air pollution investigation of 1974. All are written by scientists active in acid deposition research. All are lacking in systematic organization of the full scope of the history of acid deposition problems in Japan. All are descriptive, and none contain a theoretical analysis of historical changes.

Chapter 4

1. There are literally hundreds of works that describe the Meiji Restoration and the era before and after in greater or lesser detail. I recommend Nakane and Oishi (1991), an outstanding introduction to Tokugawa Japan; Michio and Urrutia (1985), which contains papers from a conference on the Meiji Restoration; Lehman (1978), a readable account of the period between 1850 and 1905 during which Japan rose to world power status; Rosenstone (1988), which contains biographies of Americans living in Japan during the Meiji period; Reischauer (1990), a classic history of Japan; and Fairbank et al. (1989), a renowned history of East Asia that puts it all in perspective.

2. The reader interested in a blow-by-blow (day-by-day, month-by-month, and year-by-year) account of events related to environmental issues and occupational hazards from 1868–1975 should consult Nobuko Iijima's (1979) impressive compendium, *Pollution Japan: Historical Chronology.*

3. Early accounts by actual participants in the events include: *Yanaka-mura Metsubōshi* (The Destruction of Yanaka Village) by Arahata Kanson, a novelist and social activist (Arahata 1963 [originally published in 1907]); the papers of Tanaka Shōzō, a politician and pioneer environmentalist who led the attempt to rectify the disaster; documents of the Furukawa Mining Company, which owned the mine; prefectural and national level investigative surveys; and, once Ashio

became a public issue, numerous newspaper accounts. There are dozens of later works in Japanese devoted exclusively to the events at Ashio, and hundreds that include long and short summaries. Detailed accounts in English include Notehelfer (1975), Stone (1975), and Shoji and Sugai (1992); an analysis of labor problems at Ashio (Nimura 1997); and an excellent biography of Tanaka Shōzō by Kenneth Strong (1995). Shorter accounts include Gresser et al. (1981, 4–10), numerous listings in Kodansha (1983), McKean (1981, 35–38), and material available from the Ashio Copper Mine Museum in the village of Ashio. A fine historical guidebook to the area around Ashio, which brings the century-old events to life is Fukawa and Kamiyama (1994). Other sources are cited in the chapter.

4. Furukawa Ichibei (1832–1903) was born into a tofu maker's family named Kimura in Kyoto. As a child he learned the tofu trade, but in 1849 he became a silk purchasing agent. In 1858 he was adopted by a Kyoto silk merchant by the name of Furukawa Tarōzaemon, and his family name became Furukawa. After the Meiji Restoration he helped establish the silk trade in Yokohama, the central port city in Japan. With the help of two European experts he established the first mechanical silk thread reeling factory in Japan. After the collapse in 1874 of the silk firm, he became the manager of a mining enterprise in northern Japan, and eventually purchased the mine.

In 1877 he acquired Ashio. With Ashio as his most productive mine he quickly became one of Japan's leading industrialists, and founded the Furukawa *zaibatsu* (conglomerate), which is still one of Japan's largest corporations. The modern computer company, Fujitsu, is a spin-off of Furukawa's original conglomerate. He was twice awarded the prestigious Imperial Order of Merit for his contributions to Japan's early industrialization efforts. He was one of the industrial geniuses of early modern Japan (the influential monthly magazine *Taiyō* in 1899 voted him one of the "twelve great men of Meiji"), though his reputation was greatly tarnished by the Ashio environmental disasters. He died in 1903. (For more information on Furukawa, see his official biography by Itsukakai (1926); Kodansha (1983, vol. 2, 372–373); and commentary in many of the references listed in the footnote above.)

5. One interesting record of the Ashio smoke problem, which was pointed out to me by researchers at the Forestry and Forest Products Research Institute, is contained in handwritten documents of observations at a meteorological station set up by the Forestry Department of the Ministry of Agriculture and Commerce in 1915 on Mt. Nantai in the Nikkō area near Ashio. (The reader may remember that Nikkō is the location of the flowery mausoleum of Tokugawa Ieyasu, founder of the Tokugawa Shogunate.) The meteorological station operated for four years from 1915 to 1918. Most of the station attendants did not note anything about Ashio in their records, but one enthusiastic attendant who served in 1917 observed anything and everything. He continually refers to smoke from the Ashio refinery flowing over the mountains into the valley below him. The smoke, he notes, would on occasion settle on famed Lake Chūzenji in the valley. He also notes that if the air was humid the smoke would press flat against the land, but

if dry would loft into the sky. He made no comments on the negative effects of the smoke.

6. I consulted the following documents to piece together the history of the Besshi Copper Mine: Gresser et al. (1981, 10–11), McKean (1981, 38–42), Kodansha (1983, vol. 1, 150), Shoji (1992, 46–49), and material from the Besshi Copper Mine Memorial Museum in Niihama city on the island of Shikoku.

7. Besides smelter smoke problems, labor problems also continued, and a second major strike occurred in 1925–1926. When the company fired the strikers, the mine's generating plant was sabotaged, and the Sumitomo family residence in Osaka was attacked. The prefectural governor stepped in to arbitrate the strike. Strike leaders were fired, and this solved the problem (Kodansha 1983, vol. 1, 150).

8. The Hitachi history in this section is derived from personal visits to the former Hitachi mine (and its excellent museum) and the following books in Japanese: Dai Entotsu Kinenhi Kensetsu Iinkai (1994a, 1994b), Hitachi Kinzoku K.K. (1994), and Hitachi Kōgyō K.K. (1991). English references to Hitachi include: Gresser et al. (1981, 11–12), McKean (1981, 38–42), and Wilkening (1996).

9. According to the leader of the residents movement at Hitachi, whenever Kuhara met or negotiated with them he always wore ordinary work clothes, treated them as equals, listened to their complaints, and never assumed an arrogant posture in negotiations.

10. The Chinese characters used for the word tobacco translate literally as "smoke grass."

11. One interesting influence of Hitachi's "tall stack" solution to air pollution problems turned up in a court case against the Osaka Alkali Company. Emissions from Osaka Alkali's stack caused crop damage in the area surrounding the company, and the company was sued. The case went to the Supreme Court, which ruled that the use of the best available pollution control technology would shield the company from liability, and referred the case back to the Osaka High Court. In 1919 the Osaka High Court ruled that the company had never used the best available technology because its stacks were only 30–40 meters high, whereas the stack of the Hitachi mine was over 150 meters (Gresser et al. 1981, 12–13).

12. There were plans to build a stack twice as high but the depression after the First World War ended this idea.

13. For a brief biography of Kaburagi and a reproduction of one of his scientific articles (*Jumoku no Jōchōseichō ni oyobosu Engai no Eikyō*, or "On the Effect of Smoke Damage on the Growth of Trees"; author's translation), see Ui (1991b, 86–94).

14. The first greening efforts, though, had occurred much earlier. In 1908 Oshima cherry tree seeds were brought from Oshima Island, a volcanic island just off the Pacific coast from Tokyo. These cherry trees had long evolved in the presence of volcanoes and so were relatively acid tolerant. They were planted

throughout the region and in time became, like the Big Stack, a symbol of Hitachi city.

15. To place copper mining in Japan in global and historical perspective, see the analysis of copper concentrations in Greenland ice cores and estimates of copper production over the past seven millennia in Hong et al. (1996). Hong et al.'s research does not mention Japanese copper mining per se; however, it does discuss copper mining in Song dynasty China (960–1279).

16. Damage from smelter emissions was clearly recorded and documented in early modern Japan, but damage from power plant emissions (the largest source of acid precursor emissions today) was not. The reasons for this are fairly simple. Over and above the fact that electricity generation and use was not widespread in early modern Japan (the first electric lighting in Tokyo appeared in 1886), metal ores contain much higher percentages of sulfur than fossil fuels. Thus, the amount of sulfur dioxide released is much greater from smelters than from power plants. However, in our present global energy/industrial mix, greater quantities of coal and other fossil fuels are combusted for power generation than ore smelted for metal extraction. Thus, the total amount of sulfur dioxide released from power plants today (particularly coal-fired power plants) is much greater than that released by smelters.

17. The central slogan of the day was *Fukoku Kyōhei Shokusankōgyō* (Rich Nation, Strong Military, Productive Industry).

18. The original name of today's Tokyo University was Tokyo Imperial University (*Tokyo Teikoku Daigaku*). To prevent confusion, I refer to it as Tokyo University.

19. "The Tokugawa contribution to modern science was not in the realm of the intellect but in recruitment: the kinds of people, in terms of background and class origin, who had shown serious interest in science during Tokugawa times were the same kinds who came forward after 1868" (Bartholomew 1989, 4). Most early scientists came from the samurai class, which constituted only about 6–7% of the total population at the time of the Meiji Restoration (Bartholomew 1989, 265).

20. Readers interested in the early history of acid deposition science in the West should consult Gorham (1981, 1989) and Cowling (1982).

21. Air pollution control in England began with the Alkali Act of 1863, the world's first attempt at comprehensive central government control of pollution. At this time the worst air pollution effects in England were experienced around plants manufacturing alkali, sulfate of soda, and potash. The alkali plants in particular were sending massive quantities of hydrochloric acid into the air. These factories provided processed chemicals on a large scale to the critical soap, glass, and textile industries. By the 1860s they had laid waste to whole tracts of countryside. The Alkali Act was aimed at curbing the worse of their abuses. The provisions of the Act were administered by a national Alkali Inspectorate, headed by a Chief Alkali Inspector. Robert Angus Smith was appointed the first inspector. It wasn't until 1982 that the Alkali and Clean Air Inspectorate, as it was

then called, was renamed the Industrial Air Pollution Inspectorate, and in 1987, the Inspectorate of Pollution. (For further information, see McCormick [1989, 9].)

22. Rather than "coin," "employ" is probably closer to the mark. He seems to have used the term only once toward the end of his book. (Hara Hiroshi of the National Institute for Public Health in Japan and Don Munton of the University of Northern British Columbia both pointed this interesting fact out to me.)

23. See discussion of Sudbury in McCormick (1989, 154–155).

24. See Beamish (1976) for a discussion of this early work.

25. The first scientific documentation of the effects of the Trail Smelter was Katz (1939).

26. It is worth noting that Robert Angus Smith's *Air and Rain: The Beginning of a Chemical Climatology*—which could have provided an integrating framework—was actually available in Meiji Japan. The book is today in the collection of the National Library in Tokyo. (I thank Shinichi Fujita of the Central Institute for the Electric Power Industry [CRIEPI] for pointing this out to me. See also CRIEPI [1994, 12].)

27. Even though I was unable to uncover a full list of all committee members, their influence was evident in indirect references to committee work. The general bias of scientists seemed to be toward supporting the government's position. The first committee of 1897 (with eighteen members), for instance, included mining and agricultural experts (in addition to government officials such as the Minister of Justice and the Minister of Agriculture and Commerce) (Shoji and Sugai 1992, 28; Strong 1995, 96). Evidence of their influence is the "remarkable severity" of the committee's Pollution Prevention Order (Strong 1995, 96). The second committee of 1902 looked toward long-term solutions and recommended building the huge catchment basin, which while eventually effective, was the death knell for Yanaka village. The third committee of 1910 went so far as to send experts abroad in search for countermeasures and specialists to the mines to assess damage (Shoji and Sugai 1992, 48).

28. See discussion in Stone (1975).

Chapter 5

1. I thank Fujita Shinichi of the Central Research Institute of the Electric Power Industry (CRIEPI) for bringing this information to my attention.

2. Akimoto used the colorimetric method. In other words, after adding chemicals to a sample solution, he used the color of the sample to determine the pH value.

3. Information on the history of limnology in this section is taken from Kurabayashi (1994).

4. Since this time more highly acidic crater lakes have been discovered in Japan. The most acidic is Yūgama, a crater lake in Mt. Shirane in Gunma Prefecture in

central Japan. It has a pH of about 1.0 (Satake and Saijo 1974; Takano, Saitoh, and Takano 1994). (Note: the Takano et al. article is part of a special issue on the geochemistry of crater lakes in Japan.) Modern measurements of the pH of Katanuma, the lake Yoshimura discovered, give it a pH value of about 1.8.

5. Miyake graduated from Tokyo University in 1932 and after a short stint as a research assistant at Hokkaidō University (1931–1935) was given the directorship at the new Chemical Analysis Section. He completed his doctorate at Tokyo University in 1940 with a dissertation titled "The Marine Chemistry of the Western North Pacific." Biographical information on Miyake can be found in his collected papers (Geochemical Laboratory 1978).

6. See Maruyama et al. (1993) for a more detailed description of the collection method, and for a discussion of early analytical techniques as compared to those used today.

7. Biographical information on Miyake can be found in his collected papers (Geochemical Laboratory 1978).

8. Many people refer to the current transboundary air pollution (acid rain) problem in East Asia as Japan's first experience with being a victim of extraterritorial atmospheric pollution. This is not true. Nuclear fallout from weapons testing by the United States and Soviet Union was Japan's first such experience. For a detailed discussion of the "nuclear testing regime," see Soroos (1997, 83–109).

9. This is the *Lucky Dragon*'s story. The boat known as the *Lucky Dragon* was completed and launched soon after the war in 1947 as a bonito (*katsuo*) fishing boat. At 140 tons it was one of the largest such wooden boats at the time. In the first four years of its existence it was the top bonito fishing vessel in Japan. In 1953 it was refitted as a tuna (*maguro*) fishing boat and was sent on successful fishing expeditions to the South Pacific. In 1954, with a crew of twenty-three, it set out on its last fateful fishing journey. February found it in the vicinity of the Marshall Islands, and on 1 March 1954 near the Bikini Atoll. While pursuing a school of tuna just before dawn, the astonished crew all of a sudden witnessed the sky engulfed in a fireball; a fireball ablaze with a force never before seen by humankind. They witnessed the first hydrogen bomb containing some 1,000 times the explosive power of the Hiroshima bomb.

Although more than 160 km from dead center, for more than five hours after the explosion radioactive ash rained down on the crew and ship (and on some 242 native peoples who lived in the area). Within a day crew members became nauseous and developed severe headaches. Soon after, boils and then scabs appeared in their ears, and on their face and neck, and then their hair began to fall out. At full speed the *Lucky Dragon* hurtled home in fear of what had happened. On 14 March it reached home. News of the event created a huge stir. Once again Japan had fallen victim to U.S. radiation. In May irradiated rain began falling on Japan. One crew member died in September. The fate of the *Lucky Dragon* reinvigorated Japan's opposition to nuclear weapons testing. Miyake became part of this opposition movement.

Of the boat itself, it thereafter was turned into a training vessel for the Tokyo University of Fisheries until it was retired in 1967, some twenty years after first constructed. It was junked on an artificial landfill island in Tokyo Bay known as "Dream Island." However, there were still those who remembered the fate of the *Lucky Dragon*, and a citizens movement began to save her. Finally in 1976 a refurbished *Lucky Dragon* was ensconced in a special exhibition hall on Dream Island, where it rests today as a reminder of the perils of nuclear weaponry. (See Lapp (1958) for the most detailed account in English of the *Lucky Dragon*, and visit its museum in Tokyo.)

10. For a scholar doing historical research on Japanese science at this time, there is a poignancy in the painfully thin bound volumes printed on poor quality paper, which contain scientific publications during the last years of the war and immediately after the war. Whatever else one may think of the happenings of this era, one cannot help but feel respect for those scientists who struggled to carry on research in the spirit of ideal science. Often I paused to reflect on the pursuit of science and war as I read ragged old articles in the sleek modern stacks of postwar Japan's sleek modern libraries.

11. Numerous interviewees made this comment to me.

Chapter 6

1. For a closeup view of the ugliness of this industrial pollution, see George (2001) and Oiwa et al. (2001).

2. Sources I drew upon to construct Japan's environmental history between 1945 and 1973 include: Iijima (1979, 107–401), Gresser et al. (1981, 16–51), McKean (1981, 42–79), Kodansha (1983, vol. 2, 224–229 and vol. 6, 217–220), Huddle and Reich (1987), Kamioka (1987), Hashimoto (1989), Barrett and Therivel (1991, 27–49), Hoshino (1992), and Broadbent (1998, 1–133). Citations to specific material are given in the text.

3. For further information on Yokkaichi, see Gresser et al. (1981, 105–124), McKean (1981, 59–61), Huddle and Reich (1987, 51–77), and Hashimoto (1989, 9–11).

4. See Kasuga (1989, 100–102) for a discussion of Yokkaichi asthma.

5. The most detailed English account of the trial is contained in Gresser et al. (1981, 105–124); see also Hashimoto (1989, 30–31).

6. For a description of the law and its creation, see Gresser et al. (1981, 19–27), Kodansha (1983, vol. 2, 224–225), and Hashimoto (1989, 17–20).

7. See Kato (1989) for a detailed discussion of the k-value system and other methods of controlling air pollutants between 1968 and the mid-1980s.

8. See footnote 2 of this chapter for general references containing further information on events during "The Revolution." The most detailed chronology is Iijima (1979, 280–343).

9. For more information on the formation and early structure of the Environment Agency, see Gresser et al. (1981, 26–27 and 234–245).

10. The emission standards for passenger cars used the same three pollutants, the same target levels, and the same implementation dates as contained in the U.S. Clean Air Act of 1970. The Japanese standards, though, were not legislated as in the United States, but issued by an EA administrative order. The U.S. Clean Air Act called for reductions of carbon monoxide (CO), hydrocarbons (HC), and nitrogen oxides (NO_x); each to be reduced to one tenth of the 1970 level by 1975 for CO and HC, and 1976 for NO_x. What was remarkable about the U.S. law, and reflected in the Japanese law, was that reductions were demanded before the technology was available. Reductions were based on estimates of what it would take to restore air quality, not on technological availability. Essentially technological development was mandated of the auto manufacturers. Japanese auto companies saw the handwriting on the wall and, although they started on the same footing as their U.S. counterparts, soon took the lead in developing emission control technology. They met the CO standard in 1975 as mandated (U.S. technology did not meet the standard until 1981); the HC standard in 1975 as mandated (the United States not until 1978); and the NO_x standard in 1978, two years late (the United States not until 1983). By the late 1970s Japan was the world leader in vehicle emission control technology. (For more information, see Gresser et al. (1981, 268–274), Hashimoto (1989, 34–48), and Kato (1989, 220–222).)

11. After much debate and controversy a strict twenty-four-hour standard of 0.02 ppm (equivalent to an annual average of 0.01 ppm) was promulgated. The target for achieving this standard was 1982. It was much stricter than the U.S. annual average standard of 0.05 ppm, which was enacted in 1971. The strictness was in part based on scientific studies at the time, and in part on the yet emotionally charged climate surrounding environment issues in Japan. It prompted a swirl of protest from industry. However, once promulgated the response of industry was swift. The Sumitomo Chemical Company, for instance, introduced the world's first dry denitrification process at its Chiba plant in 1973, and the Tokyo Electric Power company installed the first wet denitrification process at its Miami-Yokohama plant also in 1973. Like vehicle emission control technology, Japan became, and is still today, the world leader in denitrification technology. (For more information, see Hashimoto [1989, 48–63] and Kato [1989, 218–220].)

12. The Pollution Damage Compensation Law set up a system for compensation of pollution victims. Persons suffering from certain designated pollution diseases, such as Minamata disease or Yokkaichi type of respiratory illnesses, could apply to local authorities for certification as a pollution victim. If certified, they were eligible for compensation payments from the government, including all medical expenses and disability payments at a rate dependent on factors such as the severity of the disability, age, and sex of the victim. Polluting companies had to pay the entire cost of compensation, though administrative overhead was paid by the government. Moneys for the compensation fund came from a

pollution-load levy that was calculated based on the total amount of a pollutant discharged, e.g., the total SO_2 emitted per year. The levy was not intended as an emission control strategy, but it ended up functioning as such. Pollution areas were classified into one of two regions: class 1—those where there were widespread health affects from unspecified sources entailing various symptoms, such as Yokkaichi asthma; and class 2—those where there was specific, localized health injury, such as Minamata disease. In 1978 there were twenty-three class 1 regions with 63,653 officially designated victims. (See Gresser et al. 1981, 285–323 for a complete discussion of this law.)

13. Interview with Okita Toshiichi, 22 December 1994, and Okita (1995).

14. Okita became interested in volcanic emissions while teaching in Hokkaidō in the 1950s, and analyzed gas concentrations emitted from fumaloe of a volcano near Asahikawa. At that time he wondered if it might be possible to measure volcanic emissions remotely. Much later at a conference in the United States in 1971 he heard of the Barringer technology and persuaded the Japanese firm Jasco (Japan Spectroscopic Company, Ltd.) of Tokyo to purchase the expensive equipment. In return for the tip, which allowed Jasco to be the first company in Japan to purchase the instrument, Jasco allowed Okita to use it one time for free. Remembering the Hokkaidō experience he hurried to Oshima for one day and measured the sulfur dioxide emissions from Mt. Mihara. To his later chagrin, even though he thought this type of measurement had not been carried out before, he was too busy with air pollution work at the time to publish it in a foreign journal. (Interview with Okita Toshiichi, 22 December 1994.)

15. Information in this section derives primarily from Gorham (1981, 1989) and Cowling (1982).

16. The stories of the scientists who worked to uncover the cause-and-effect mechanisms in the Minamata, itai-itai, and Yokkaichi cases are fascinating, poignant, and impressive. The case of Hagino Noboru, the central figure in the search for the cause of itai-itai disease is illustrative (Gresser et al. 1981, 56; McKean 1981, 45–48). Hagino first proclaimed in 1955 that he believed the disease was caused by heavy metals from nearby Mitsui mining operations, an allegation Mitsui totally denied. Thereafter, he began a long crusade to win acceptance for his theory and just compensation for the victims. In 1961 he identified cadmium as the heavy metal causing the disease.

Mitsui countered that the disease was due to nutritional deficiencies in the diets of the victims. Hagino came under heavy pressure to retract his results. Threatening telephone calls, verbal harassment, and social ostracization were his reward. Unable to find funds from any source in Japan, he applied for and was awarded a $30,000 grant from the U.S. National Institute of Health to continue his research. He was eventually vindicated. In 1966 the MHW officially investigated the cause and issued a report that cadmium had been identified in the waters of the Jintsū River and in the effluent from the mine, but concluded that cadmium was only a contributing factor. However, in 1968 MHW reversed its previous stance and unequivocally stated that cadmium was the cause of itai-itai disease, and officially designated it as a pollution disease. Mitsui, however, called it a "hasty conclusion hard to understand."

Chapter 7

1. The primary sources I used to construct my image of the fourth environmental era were: McKean (1981), Huddle and Reich (1987, 1–24), Hashimoto (1989), Barrett and Therivel (1991, 38–47), 92 NGO Forum Japan (1992, 78–80), and OECD (1994).

2. See Barrett and Therivel (1991, 171–193) for an extended discussion of the Seto Inland Sea Bridge and Kansai International Airport projects.

3. See Huddle and Reich (1987, 14–23) for a discussion of NGO and citizen activities during the 1980s.

4. The National Institute for Environmental Studies (NIES) is the research arm of the Environment Agency. It is located in Tsukuba near Tokyo. Initially, its focus was pollution control, but it underwent a major reorganization in 1990 and now also devotes attention to global environmental problems and the ecological and social aspects of environmental problems. There are two major project divisions (Global Environment and Regional Environment) and six research divisions (Social and Environmental Systems, Environmental Chemistry, Environmental Health Sciences, Atmospheric Environment, Water and Soil Environment, and Environmental Biology). As of 1995 NIES had about 250 full-time researchers and a budget of about US$70 million. Scientists are organized into research teams within each division. The Acid Deposition Research Team, headed by Satake Kenichi, is housed under the Global Environment Division. I was located at NIES for two years while conducting much of the research contained in this book. (For further information on NIES, visit its web site at <http://www.nies.go.jp>.)

5. See Hashimoto (1989, 34) and Kato (1989, 213–216 and 218–220) for discussions of the total mass emission control system.

6. Okita believes he was chosen to head the committee because, first of all, he had extensive experience with air chemistry, particularly sulfate compounds; second he was a section director within a national laboratory; and third he had worked with the EA on many occasions and they knew him well (interview with Okita, 22 December 1994). Establishment of the EA's National Institute for Environmental Studies (NIES) had just been authorized in 1974 and was not yet functioning. Therefore, the EA had to call on outside help to investigate a problem like moist air pollution. Okita subsequently moved to NIES. He tells of his lifelong research on air pollution problems in Okita (1995).

7. Okita reports that he initially learned of the European acid rain problem in the mass media. This then sent him looking for existing literature. (Interview with Okita, 22 December 1994.)

8. Aldehydes are a class of chemical compounds represented by the general formula $R \cdot CHO$, where R is a radical attached to a CHO group. Aldehydes are

highly chemically reactive. This is why their presence is cause for medical concern. The simplest aldehyde is formaldehyde, which has the formula H·CHO. Acrolein, or acrylic aldehyde, is another of the numerous aldehydes, and is represented by the formula CH_2=CH·CHO.

9. There was good reason for Yoshitake being circumspect. Yoshitake had a great deal of trouble getting his results published, and when they finally were published it was in significantly altered form. The source of the difficulty was bureaucrats in the Ministry of Agriculture, Forestry, and Fisheries (MAFF) who did not want research results published that further implicated industrial development as a cause of environmental damage. The Japanese bureaucracy was still, in the late 1970s, in the throes of reacting to the explosion of citizen wrath against unmitigated industrial expansion, and so did not want to jeopardized continued development at Tomakomai in Hokkaidō with negative findings.

In addition, by the late 1970s the MAFF had taken a stand that there was no forest damage due to acid rain in Japan. And then all of a sudden it was faced with evidence that there was indeed such damage in Japan! The first Yoshitake paper (Yoshitake and Hisao 1978) was delayed a year in publication. The second paper (1980) contains no reference to "acid rain"—the authors were told not to use the term.

The third paper (Yoshitake and Hisao 1986) was not only delayed years, but the authors were forced to radically alter their original conclusion, i.e., that the forest damage was due to acid fog from Tomakomai industrial sites, especially a large pulp and paper mill directly upwind of the damage area. The title of the original draft of the paper contained the words "acid fog." This was deleted. A map in the original contained the location of the pulp and paper mill. Its location was deleted. A schematic diagram of the set of processes from emissions to acid fog formation to tree damage was included in the original. It was eliminated. The conclusion of the original version was only vaguely hinted at in the 'revised' version. Until the paper met the approval of MAFF bureaucrats, it was not cleared for publication. This process took six years. By the time the 'revised' paper was published in 1986 it coincided with that of Sekiguchi's more controversial paper. Sekiguchi directly pinned blame for a decline of Japanese cedar in the Kantō area to acid rain.

Sekiguchi's paper triggered widespread news coverage and brought the issue of acid rain and forest damage to the Japanese public's attention for the first time. At this time neither Yoshitake nor Sekiguchi were aware of the other's work. Yoshitake's work was more scientifically rigorous, but Sekiguchi's more provocative and attention-grabbing. Thus, Sekiguchi's work is often referenced, but Yoshitake's seldom is. However, as we shall see in the next chapter, Sekiguchi was forced to publish in a foreign journal, and the controversy he provoked succeeded in getting him banished from acid rain research. The experiences of these two researchers well illustrate the perils of researchers in Japan who venture or stumble into the realm of controversial environmental research. (The preceding information derives from interviews with participants in the events described.)

10. This conclusion is based on comments by several interviewees who participated in moist air pollution research.

11. Literally hundreds of works describe European and North American acid rain science and politics during the 1970s and 1980s. For further information on Europe, see Bolin (1971) and Brosset (1973) for early accounts of Sweden's acid rain case study presented at the 1972 Stockholm Conference; Boehmer-Christiansen and Skea (1991) for a comparative study of British and German acid rain politics; and Sand (1987) for European international air pollution policy. For further information on North America, see Likens et al. (1972), Cogbill and Likens (1974), and Cogbill (1976) for accounts by the "discoverers" of acid rain in North America; Summers and Whelpdale (1976) for a description of acid rain in Canada; and Schmandt et al. (1988) for a comprehensive overview of U.S.-Canadian acid rain relations. For general works that cover both Europe and North America, see Weller (1980), Ashby and Anderson (1981), Wetstone and Rosencranz (1983), Gould (1985), Regens and Rycroft (1988), "the case of acid rain" in Carroll (1988), McCormick (1989), which gives a global overview, and Munton et al. (1999). Also see the previously cited histories of early acid deposition research by Cowling (1982) and Gorham (1981).

12. See Overrein (1976) for a description of SNSF, and Overrein et al. (1980) for the final report.

13. See Jackson (1990) for an account of the events leading to the signing of the convention.

14. See Pearce (1986) for an overview of *Waldsterben*.

15. Seliga and Dochinger (1976, 135).

Chapter 8

1. The English name of the survey has changed since its inception in 1983. It was first called the Acid Rain Survey. This was changed to Acid Precipitation Survey in the late 1980s, and then switched to Acid Deposition Survey. The Japanese title, however, has remained the same.

2. Interview with Kato Saburo, 10 April 1996.

3. Japan is such a narrow country that when considering domestic pollutant transport generally only medium-range (hundreds of kilometers), not long-range (1,000 or more kilometers), distances are relevant. Thus, in this section I refer to early domestic analyses and modeling efforts as medium-range or meso-scale. However, most of the early literature from the 1980s refers to this as "long-range transport." I have reserved the term "long-range transport" solely for transport between Japan and the Asian mainland. Under this definition long-range transport models in Japan don't show up until the late 1980s, and don't really begin proliferating until the early 1990s.

4. They were published in more detail than contained in the Final Report in the following articles: Hara et al. (1990); Kitamura et al. (1991); Mori et al. (1991); Tamaki et al. (1991).

5. Nonseasalt sulfate (nss-SO_4^{2-}): It is common practice in precipitation chemistry analysis to distinguish between sulfate ions that originate from oceans (seasalt sulfate—ss-SO_4^{2-}) and those that originate from all other sources (nonseasalt sulfate—nss-SO_4^{2-}). Nss-SO_4^{2-} is generally estimated by assuming that all sodium ions (Na^+) derive from seawater, and that the ratio of Na^+ to ss-SO_4^{2-} in precipitation is identical to that in seawater. Thus, if the Na^+ concentration is known in precipitation, then the ss-SO_4^{2-} fraction of total sulfate concentration can be calculated, and thus, the nss-SO_4^{2-} fraction determined by simple subtraction.

6. Precipitation rates are far higher in Japan than North America, 1,800 mm per year versus about 760 mm on the northeast coast of North America. Thus, since deposition rates are defined as the product of ionic concentrations in precipitation times precipitation amount, an equivalent deposition rate for Japan and North America means that ionic concentrations in precipitation are smaller for Japan.

7. The ad I saw appeared in the 31 August 1994 issue of the newspaper *Yomiuri Shimbun*, page 28. Even though this example is from the 1990s, it is indicative of what one could find in the 1980s.

8. Uemura Shinsaka (1993) of Osaka, for instance, started the so-called Morning Glory Acid Rain Survey. In 1987 he began experimenting with the use of morning glories as a means of observing acid rain at home. He was interested in morning glories because the very first reports of acid rain damage in Japan were reports of bleaching and spotting on morning glory petals, because morning glory seeds were easy to grow, and because the petals were relatively sensitive to acid. After testing his morning glory method, in 1989 he distributed 350 morning glory sets. Out of 189 people who reported results, 163 observed spots. The campaign caught on and in 1990 it was touted on TV, radio, and in newspapers, and 500 sets were distributed nationwide. He also gave out a thirty-two-page pamphlet of his own creation titled "Morning Glories are Crying: An Introduction to the Acid Rain Problem" in 1990. The survey provoked an enthusiastic response, and inspired similar uses of morning glories in environmental fairs, school classrooms, and citizen environmental education programs.

9. For more detailed discussion of rain, forests, Japanese culture, and citizen acid rain activities, see Wilkening (1999).

10. See Levy (1993) and Hordijk (1995) for discussions of these entities.

11. See Swedish NGO Secretariat on Acid Rain (1995) for an introduction to the critical loads concept.

12. See Swedish NGO Secretariat on Acid Rain (1995) for a nice summary of how critical loads are calculated.

Chapter 9

1. For further information on Japan's international environmental activities before and during the Global Environment Era, see Gresser et al. (1981,

353–385), Forrest (1986), Murdo (1990, 1992), Miller and Moore (1991), EA (1992, 602–640), 92 NGO Forum Japan (1992, 87–108), Imura (1994), Schreurs (1994), Tsuru (1999, 116–139 and 193–223), and Wong (2001).

2. "The World's Eco-Outlaw?" *Newsweek*, 1 May 1989, 70.

3. Cross (1988, 38).

4. Murdo (1990, 3).

5. See the appendix for a detailed discussion of the present state of science in Japan related to the international dimension of its acid deposition problem.

6. This and other models as they relate to global environmental problems are discussed in Schreurs (1994) and Wong (2001).

7. For discussions of the role of the bureaucracy in Japanese politics, see Gresser et al.'s *Environmental Law in Japan* (1981), Chalmers Johnson's *MITI and the Japanese Miracle* (1982), and Kōichi Kishimoto's *Politics in Modern Japan* (1988).

8. See Wong (2001, 231–239) for a discussion of the battle between the EA and MITI for control of the international acid deposition issue.

9. MITI did, however, attempt to establish an international monitoring network but the effort fizzled.

10. See Wilkening (1999) for further information.

11. Although it is hard to derive an exact figure, this estimate is based on the rough average of estimates given by various leading researchers who were asked in interviews to estimate the size of the community, and by the number of individuals who are members of acid deposition-related sections within several of Japan's scientific societies. For example, the membership list of the Acid Deposition Section (*Sanseiu Bunkakai*) of the Atmospheric Environment Society of Japan (*Nihon Taiki Kankyō Gakkai*) contained some 200 individuals in 1995.

12. The sixth International Conference on Acid Deposition, or Acid Rain 2000, was the latest in a series of international conferences on acid deposition science and policy held once every five years. The series began in 1975. Acid Rain 2000 was the first to be held outside of Europe or North America.

13. In a February 1984 meeting with Japanese officials, William Ruckelshaus, head of the U.S. EPA, raised the issue of long-range transport of air pollution in East Asia. Schreurs et al. (1995, G10), from whose work this piece of information was drawn, report that eight out of seventeen articles on acid rain appearing in *Asahi Shimbun* in 1985 mention the acid rain problem in China and the potential of long-range transport of sulfur dioxide to Japan.

14. For instance, Kitamura Moritsugu of the Ishikawa Prefecture Environmental Research Center reported such findings in 1986. For a discussion of this and other early reports, see Bi et al. (1994) and Hara (1993).

15. For a discussion of Japan's early involvement in the global climate change and stratospheric ozone depletion issues, see Schreurs (1994).

16. See, for instance, *Christian Science Monitor*, 6 November 1992, 1—"Japanese Link Increased Acid Rain to Distant Coal Plants in China."

17. MITI's "Green Aid Plan" was basically an attempt to transfer and adapt Japan's successful experience with combating its own industrial pollution problems to select developing countries. The four environmental areas targeted were water pollution, air pollution, waste management and recycling, and energy efficiency and energy substitution. In each of these areas surveys of technological needs, transfers of technology, cooperative research and development of new technology, training programs, and demonstration projects were conducted.

In the first year of its implementation, 1992, China and Thailand were the focus countries. In 1993, Indonesia was added to the list, in 1994 the Philippines and Malaysia were added, and in 1995 India was incorporated into the plan. Funding began at U.S.$100 million in 1992, and was expanded to U.S.$120 million in 1993 and U.S.$140 million in 1994. For more information on Japan's efforts to develop and disseminate environmental technology, see Moore and Miller's (1994) account of the race between Japan, Germany, and the United States to develop green technologies. See also Myers (1992) and Jubak and D'Amico (1993).

18. In 1993 a multinational, cooperative project initiated by Western scientists and funded by the World Bank and the Asian Development Bank began a study of the impacts of acid deposition in Asia. Known as the "Acid Rain and Emission Reductions in Asia," or Rains-Asia project, its primary objective was to develop a policy-oriented integrated assessment computer model—RAINS-Asia—patterned after the European RAINS model (see the previous chapter for a discussion of the RAINS model). Phase 1 of the project was completed in June 1995, and a final report issued (Foell et al. 1995). For more information on the RAINS-Asia Project, see the project homepage at <http://www.iiasa.ac.at/Research/TAP/rains_asia/docs/>, visited 1 July 2001.

19. The project found that the corrosion rates in China were significantly higher than those in Japan and South Korea. For instance, the rates of corrosion of bronze in Chongqing (6.81 micrograms per year), which had the highest corrosion rate in China, were about three times greater than Tokyo (1.98), which had the highest rate in Japan, and eight times greater than Kyoto (0.84) (Maeda 1995). The rates for other metals and stone showed similar ratios. In addition, corrosion rates in Ishikawa Prefecture on the Sea of Japan coast were about four times higher in winter than in summer. The project leader stated: "It is most likely that this is due to sulfur dioxide emissions from the continent being converted to sulfate, carried on winter winds to Japan, and incorporated into snowfall on the Sea of Japan coast. Therefore, Japan cannot afford to overlook China's air pollution problems" (quoted in *Asahi Shimbun* 10 November 1995, 19; author's translation).

20. For more information on the state of the acid deposition problem in China in the late 1980s and early 1990s, see Zhao and Sun (1986a, 1986b), Galloway et al. (1987), Zhao and Xiong (1988), Zhao et al. (1988), Quan (1991), State

Council of China (1992), Quan (1993), Wang (1993), Wang et al. (1993), Cheng (1993), Feng and Ogura (1993), Bi et al. (1994), Zhao (1994), Huang et al. (1995), Qi et al. (1995), Seip et al. (1995), and Wang and Wang (1995). I interviewed Wang Rusong of the Chinese Academy of Sciences, 11 December 1996.

21. See the country reports from Korea in EA (1993). Also see MOE (1995, 12 and 27–34). I interviewed Lee Dong Soo of Yonsei University, 14 February 1996; Nam Jae-Chol of the Meteorological Research Institute, 15 February 1996; Kim Man-Goo of Kangwon University, 15 February 1996; and Park Soon-Ung of Seoul National University, 10 December 1996.

22. On a couple of occasions I had a chance to talk with energy experts from North Korea and asked them about acid rain in their country. They told me that there were a handful (about five) monitoring stations around the country, but the equipment was old and unreliable. Since emission controls were nonexistent, they speculated that acid rain was the cause of difficulties experienced by reforestation projects. Seedlings soon died. They were very much surprised when I told them that long-range transport from China might be affecting their country. They had never heard of transboundary transport in the region.

23. See Liu et al. (1993) and EPA-Taiwan (1993, 34–93). I interviewed Wu Yee-Lin of National Cheng Kung University, 14 February 1996.

24. See the country report from Russia in EA (1993, 286–300) and Ryaboshapko et al. (1994).

25. See the country report from Mongolia in EA (1993, 248–250).

26. For more information on the history and structure of EANET, see the network homepage at <http://www.adorc.gr.jp/>, visited 1 July 2001. See also EANET information on the Ministry of Environment web site <http://www.env.go.jp/en/topic/eanet/ch1.html>, visited 1 July 2001.

27. This conclusion comes from discussions with Japanese and Chinese participants in a workshop "Innovative Financing of Clean Coal in China: A GEF Technology Risk Guarantee?" held 27–28 February 1999 in Berkeley, California.

28. The history and structure of the network to the mid-1990s is discussed in Yagishita (1995).

29. The author served on the Environment Agency Secretariat for the second and third Expert Meetings in a translation and English editing capacity.

30. The results of the First and Second Meetings of the Working Group on the Acid Deposition Monitoring Network in East Asia, and the First Intergovernmental Meeting on the Acid Deposition Monitoring Network in East Asia are contained in Interim Secretariat of EANET (1998a).

31. The results of the Third Meeting of the Working Group, and the First Interim Scientific Advisory Group are contained in Interim Secretariat of EANET (1998b).

32. See <http://www.adorc.gr.jp/event/ig02.html> for all documents produced by the Second Intergovernmental Meeting, visited 1 July 2001.

33. "Joint Announcement on the Implementation of the Acid Deposition Monitoring Network in East Asia (EANET)," <http://www.adorc.gr.jp/event/ig02/ig2_5_2.pdf>, visited 1 July 2001.

34. An early evaluation of the applicability of the critical loads approach in Japan is Shindo et al. (1995).

Appendix

1. An excellent summary of the framework as it stood in the mid-1990s can be gained by the papers by Japanese authors given at the International Symposium on Acidic Deposition and its Impacts, 10–12 December 1996, Tsukuba, Japan (NIES 1996).

2. GEPRP background information, budgets, and brief descriptions of projects are contained in the yearly government document put out by the EA, *Chikyū Kankyō Hozen Chōsa Kenkyū nado Sōgō Suishin Keikaku Jisshi Jōkyō Hōkokusho*, which essentially translates as the GEPRP annual report.

3. The author was responsible for the English editing of this document, which was published just in time for distribution at the fifth International Conference on Acid Deposition held in Sweden in June 1995.

4. For more information on CRIEPI's acid deposition project, see the following major reports CRIEPI (1992, 1994) and Kohno (1997). Besides numerous visits to CRIEPI's Komae Research Laboratory, I interviewed Ichikawa Yoichi, 27 July 1995; Sato Kazuo, 27 July 1995; and Fujita Shinichi, 22 March 1996.

5. For background information on local government environmental institute work, see Acid Precipitation Survey Section of the Association of Local Government Environmental Research Institutes (1995). I also interviewed Tamaki Motonori of the Hyogo Prefectural Institute for Environmental Studies on 3 August 1995. Tamaki was one of the leaders of the Acid Precipitation Survey Section.

6. Interview with Kaneyasu Naoki, 19 June 1995.

7. GO3OS began in 1957 as part of the International Geophysical Year (IGY) and is now a network of 120 stations around the world that record ozone data. Data provided by the network is a crucial input into the negotiation and implementation of the Vienna Convention for the Protection of the Ozone Layer and its protocols. Although ozone, particularly tropospheric ozone, is relevant to the acid deposition problem, the data provided by the GO3OS network are not as directly applicable to the acid deposition problem as those of the BAPMoN network, thus the GO3OS network is not discussed here.

8. Information on GO3OS and BAPMoN is from JMA (1994, 126–129).

9. Information about the survey is from Acid Precipitation Survey Section of the Association of Local Government Environmental Research Institutes (1995).

10. Domestically, the EA compiles nationwide inventories for SO_2, NO_x, and PM from stationary sources every year, starting in 1978. In 1995, of a total of

710,000 tonnes of SO_2, 26% originated from electric power plants, 13% from the chemical industry, 11% from the steel industry, and the rest from a variety of other sources. The EA also compiles NO_x, HC, and PM inventories for motor vehicle emissions. In 1994, of a total of 550,000 tonnes of NO_x, 75% was emitted from diesel engines and 25% from gasoline engines. The preceding data is from Murano et al.(1999).

11. The author worked part-time at the consulting firm Resource Management Associates (RMA) of Madison, Wisconsin, for about two years from 1991–1993 assisting in the development of the RAINS-Asia project emission inventory.

12. Interview with Katatani Noritaka of Yamanashi University, 8 August 1995.

13. The most comprehensive summary of the ecological impacts of acid deposition in Japan is Satake (1999).

References

Note: *Denotes that the English title is my translation; otherwise English titles are reproduced as contained in the original.

92 NGO Forum Japan. 1992. *People's Voice of Japan*. Tokyo: 92 NGO Forum Japan.

Acid Precipitation Survey Section of the Association of Local Government Environmental Research Institutes. 1995. Sanseiu Zenkoku Chōsa Kekka Hōkokusho (Nationwide Acid Precipitation Survey Final Report).* *Zenkoku Kōgai Kenkaishi (Journal of the Environmental Laboratories Association)* 20(2): 1–74.

AirNote. (no date, 1972[?]). Volcanic SO_2. San Francisco: Environmental Measurements, Inc. of San Francisco, CA.

Akimoto, Hajime, and Hirohito Narita. 1994. Distribution of SO_2, NO_x and CO_2 Emissions from Fuel Combustion and Industrial Activities in Asia with $1° \times 1°$ Resolution. *Atmospheric Environment* 28(2): 213–225.

Akimoto, Minoru. 1929. Osaka niokeru Usui no Seijō nitsuite (On the Nature of Rainfall in Osaka).* *Kokumin Eisei (Public Health)** 6(4): 463–470.

Alcamo, Joseph, R. W. Shaw, and Leen Hordijk. 1990. *The RAINS Model of Acidification: Science and Strategies in Europe*. Dordrecht, The Netherlands: Kluwer Academic Publishers.

Allison, Graham T., and Morton H. Halperin. 1972. Bureaucratic Politics: A Paradigm and Some Policy Implications. In *Theory and Policy in International Relations*, eds. Raymond Tanter and Richard H. Ullman, 40–79. Princeton: Princeton University Press.

Applied Meteorology Research Department. 1995. *Io Sankabutsu no Chōkyori Usō Moderu to Higashi Ajia Chiiki e no Tekiyō (The Long-range Transport Model of Sulfur Oxides and Its Application to the East Asia Region), Technical Report No. 34*. Tsukuba, Japan: Meteorological Research Institute.

Arahata, Kanson. (1963 [originally published in 1907]). *Yanaka Metsubōshi (The Destruction of Yanaka Village).** Tokyo: Meiji Bunken.

Asahi Shimbun. 1993a. Nihon e Ekkyō Osen no Osore, Chūgoku de SO2 Haishutsuzō (Fear of Transboundary Pollution to Japan, Increasing SO2 Emissions in China).* *Asahi Shimbun* (24 January), 1.

Asahi Shimbun. 1993b. Kansoku Kenkyū, Nihon Omoi Yakuwari (Research and Monitoring, Japan's Important Role).* *Asahi Shimbun* (25 February), 5.

Ashby, Eric, and Mary Anderson. 1981. *The Politics of Clean Air*. Oxford, UK: Clarendon Press.

Asquith, Pamela J., and Arne Kalland, eds. 1997. *Japanese Images of Nature: Cultural Perspectives*. Richmond, UK: Curzon Press.

Barrett, Brendan F. D., and Riki Therivel. 1991. *Environmental Policy and Impact Assessment in Japan*. London: Routledge.

Barrie, Leonard A. 1986. Arctic Air Pollution: An Overview of Current Knowledge. *Atmospheric Environment* 20(4): 643–663.

Bartholomew, James R. 1989. *The Formation of Science in Japan: Building a Research Tradition*. New Haven, CT: Yale University Press.

Beamish, Robert J. 1976. Acidification of Lakes in Canada by Acid Precipitation and the Resulting Effects on Fishes. *Water, Air, and Soil Pollution* 6(2/3/4): 501–514.

Bi, Tong, Shijun Qiao, and Moritsugu Kitamura. 1994. Chūgoku niokeru Sanseiu to sono Taisaku nitsuite (On the State of Acid Rain and Acid Rain Policy in China).* *Kankyō Gijutsu (Environmental Conservation Engineering)* 23(11): 666–669.

Boehmer-Christiansen, Sonja, and Jim Skea. 1991. *Acid Politics: Environmental and Energy Policies in Britain and Germany*. London: Belhaven.

Bolin, Bert, ed. 1971. *Air Pollution Across National Boundaries: The Impact on the Environment of Sulfur in Air and Precipitation*. Stockholm: Norstedt & Söner.

Brecher, Puck W. 2000. *An Investigation of Japan's Relationship to Nature and Environment*. Lewiston, NY: Edwin Mellen Press.

Broadbent, Jeffrey. 1998. *Environmental Politics in Japan: Networks of Power and Protest*. New York: Cambridge University Press.

Brosset, Cyrill. 1973. Air-Borne Acid. *Ambio* 2(1–2): 2–9.

Brundtland, Gro Harlem. 1997. The Scientific Underpinning of Policy. *Science* 277: 457.

Brydges, T. G., and R. B. Wilson. 1991. Acid Rain Since 1985—Times Are Changing. In *Acidic Deposition: Its Nature and Impacts*, eds. F. T. Last and R. Watling, 1–16. Edinburgh, Scotland: Royal Society of Edinburgh.

Bubenick, David V., ed. 1984. *Acid Rain Information Book*, 2nd edition. Park Ridge, NJ: Noyes Publications.

Caltabiano, Tommaso, Romolo Romano, and Gennaro Budetta. 1994. SO_2 Flux Measurements at Mount Etna (Sicily). *Journal of Geophysical Research* 99(D6): 12,809–12,819.

Carmichael, Gregory R., Leonard K. Peters, and Toshihiro Kitada. 1986. A Second Generation Model for Regional-Scale Transport/Chemistry/Deposition. *Atmospheric Environment* 20(1): 173–188.

Carmichael, Gregory R., Leonard K. Peters, and Rick D. Saylor. 1991. The STEM-II Regional Scale Acid Deposition and Photochemical Oxidant Model—I. An Overview of Model Development and Applications. *Atmospheric Environment* 25(10): 2077–2090.

Carroll, John E., ed. 1988. *International Environmental Diplomacy: The Management and Resolution of Transfrontier Environmental Problems*. Cambridge, UK: Cambridge University Press.

Chang, J. S., R. A. Brost, I. S. A. Isaksen, S. Madronich, P. Middleton, W. R. Stockwell, and C. J. Walcek. 1987. A Three-Dimensional Eulerian Acid Deposition Model: Physical Concepts and Formulation. *Journal of Geophysical Research* 92(D12): 14,681–14,700.

Chang, YoungSoo, Gregory R. Carmichael, Hidemi Kurita, and Hiromasa Ueda. 1989a. The Transport and Formation of Photochemical Oxidants in Central Japan. *Atmospheric Environment* 23(2): 363–393.

———. 1989b. The Transport and Formation of Sulfates and Nitrates in Central Japan. *Atmospheric Environment* 23(8): 1749–1773.

Chang, YoungSoo, B. S. Ravishanker, Gregory R. Carmichael, Hidemi Kurita, and Hiromasa Ueda. 1990. Acid Deposition in Central Japan. *Atmospheric Environment* 24(8): 2035–2049.

Cheng, Zifeng. 1993. State and Trend of Acid Precipitation Monitoring in China. In *First Expert Meeting on Acid Precipitation Monitoring Network in East Asia (26–28 October), Toyama, Japan*, 187–195. Tokyo: Environment Agency of Japan.

Cogbill, Charles V. 1976. The History and Character of Acid Precipitation in Eastern North America. *Water, Air, and Soil Pollution* 6(2/3/4): 407–413.

Cogbill, Charles V., and Gene E. Likens. 1974. Acid Precipitation in the Northeastern United States. *Water Resources Research* 10(6): 1133–1137.

Cowling, Ellis B. 1982. Acid Precipitation in Historical Perspective. *Environmental Science & Technology* 16(2): 110A–123A.

CRIEPI (Central Research Institute of the Electric Power Industry). 1985. *Data Report on the Inland Sea Regional Acid Deposition Project*. Tokyo: CRIEPI.

———. 1992. *Acidic Deposition in Japan, Report # ET91005*. Tokyo: CRIEPI.

———. 1994. *Sanseiu no Eikyō Hyōka (Evaluation of the Effect of Acid Deposition).** Tokyo: CRIEPI.

Cross, Michael. 1988. Japan Wakes up to the Environment. *New Scientist* 118: 38–39.

Dai Entotsu Kinenhi Kensetsu Iinkai (Committee for Construction of a Memorial to the Big Stack).* 1994a. *Dai Entotsu Kinenhi Setsuritsu (Commemoration of a Memorial to the Big Stack).** Hitachi: Hitachi Shimin Bunka Jigyōdan.

———. 1994b. *Hitachi Dōzan Engai Mondai Mukashibanashi (Story of the Hitachi Copper Mine's Smoke Problem).** Hitachi: Hitachi Shimin Bunka Jigyōdan.

EA (Environment Agency of Japan). 1987a. *Sanseiu Taisaku Chōsa Chūkan Hōkokusho (Interim Report of the Acid Rain Survey).** Tokyo: Environment Agency of Japan.

———. 1987b. *Interim Report of Acid Precipitation Survey in Japan.* Tokyo: Environment Agency of Japan.

———. 1989. *Dai Ichi Ji Sanseiu Taisaku Chōsa Kekka (Results of the Acid Precipitation Survey: Phase 1).** Tokyo: Environment Agency of Japan.

———. 1990. *Acid Precipitation in Japan: The Report of Phase I Survey.* Tokyo: Environment Agency of Japan.

———. 1992. *Quality of the Environment in Japan, 1992.* Tokyo: Environment Agency of Japan.

———. 1993. *The Expert Meeting on Acid Precipitation Monitoring Network in East Asia (26–28 October), Toyama, Japan.* Tokyo: OECC (Overseas Environmental Cooperation Center).

———. 1994. *Dai Ni Ji Sanseiu Taisaku Chōsa Kekka (Results of the Acid Precipitation Survey: Phase 2).** Tokyo: Environment Agency of Japan.

———. 1995a. *The Second Expert Meeting on Acid Precipitation Monitoring Network in East Asia (22–23 March), Tokyo.* Tokyo: OECC (Overseas Environmental Cooperation Center).

———. 1995b. *Final Report of the Third Expert Meeting on Acid Deposition Monitoring Network in East Asia (14–16 November), Niigata, Japan.* Tokyo: OECC (Overseas Environmental Cooperation Center).

———. 1995c. *Acid Deposition Survey: Phase 2 (1988–1993) Final Report.* Tokyo: Environment Agency of Japan.

———. 1997a. *Acid Deposition Monitoring Network in East Asia: Achievements of the Expert Meetings.* Tokyo: Environment Agency of Japan.

———. 1997b. *Guidelines and Technical Manuals for Acid Deposition Monitoring Network in East Asia: Adopted by the Expert Meetings on Acid Deposition Monitoring Network in East Asia.* Tokyo: Environment Agency of Japan.

———. 1999. *Dai San Ji Sanseiu Taisaku Chōsa Torimatome (Results of the Acid Deposition Survey: Phase 3)**. Tokyo: Environment Agency of Japan.

Easton, David. 1953. *The Political System.* New York: Knopf.

Efinger, Manfred, Peter Mayer, and Gudrun Schwarzer. 1993. Integrating and Contextualizing Hypotheses. In *Regime Theory and International Relations*, ed. Volker Rittberger, 252–281. Oxford: Clarendon Press.

EPA-Taiwan (Environmental Protection Administration of the Republic of China). 1993. *State of the Environment, Taiwan, R.O.C, 1993.* Taipei: Environmental Protection Administration of the Republic of China.

ESCAP (Economic and Social Commission for Asia and the Pacific). 1999. *Proceedings of the Expert Group Meeting on Emission Monitoring and Estimation.* Tokyo: Environment Agency of Japan.

Fairbank, John K., Edwin O. Reischauer, and Albert M. Craig. 1989. *East Asia: Tradition and Transformation (3rd edition)*. Boston: Houghton Mifflin.

Feng, Zongwei, and Norio Ogura, eds. 1993. *Proceedings of the China–Japan Joint Symposium on the Impacts and Control Strategies of Acid Deposition on Terrestrial Ecosystems, 1–4 November 1992, Beijing*. Beijing: China Sciences & Technology Press.

Foell, Wes, Markus Amann, Greg Carmichael, Michael Chadwick, Jean-Paul Hettelingh, Leen Hordijk, and Zhao Dianwu, eds. 1995. *RAINS-ASIA: An Assessment Model for Air Pollution in Asia*. Washington, DC: World Bank.

Forrest, Richard A. 1986. Kogai to Gaiko: Japan and the World Environment (Master's thesis, University of Michigan, Ann Arbor).

Forster, Bruce A. 1993. *The Acid Rain Debate: Science and Special Interests in Policy Formation*. Ames, IA: Iowa State University Press.

Foucault, Michel. 1970. *The Order of Things: An Archeology of the Human Sciences*. Translated by A. M. S. Smith. London: Travistock/Routledge.

———. 1972. *The Archaeology of Knowledge*. Translated by A. M. S. Smith. New York: Harper & Row.

Fujii, Tetsu. 1971. Asagao no Kaben no Dasshoku (Discoloration of Morning Glory Petals).* *Taiki Osen Nyūsu (Air Pollution News Report)* 62(1): 1.

Fujimura, Joan H. 1992. Crafting Science: Standardized Packages, Boundary Objects, and "Translation." In *Science as Practice and Culture*, ed. Andrew Pickering, 168–211. Chicago: University of Chicago Press.

———. 1996. *Crafting Science: A Sociohistory of the Quest for the Genetics of Cancer*. Cambridge, MA: Harvard University Press.

Fujita, Shinichi. 1987. Wagakuni no Genjō (The State of Acid Rain in Japan).* *Kishō Kenkyū Nooto (Meteorological Research Notes)** 158: 23–35.

———. 1990. Kyūshū Hokusei Kaiiki niokeru Iōkagōbutsu no Nōdo to Chinchakuryō (Chemical Composition of Precipitation and Wet and Dry Deposition of Sulfur Compounds around the Northwestern Kyushu Region). *Taiki Osen Gakkaishi (Journal of Japanese Society of Air Pollution)* 25(2): 155–162.

———. 1993a. Sanseiu Kenkyū Hyakunen no Rekishi to sono Hensen: 1872 Nen kara 1972 Nen no Kōsui no Kagaku, Dai 1 Kai to Dai 2 Kai (100 Years of Acid Rain History and its Changes: Precipitation Chemistry from 1872 to 1972, Part 1 and Part 2).* *Shigen Kankyō Taisaku (Journal of Resources and Environment)* 29(7 & 8): 69–75, 82–88.

———. 1993b. Kazan Katsudō to Kankyō no Sanseika (Volcanic Activity: Some Effects of the Emissions on the Acidification of the Environment). *Taiki Osen Gakkaishi (Journal of Japanese Society of Air Pollution)* 28(2): 72–90.

———. 1997. Overview of Acidic Deposition Assessment Program in CRIEPI. In *Proceedings of CRIEPI International Seminar on Transport and Effects of Acidic Substances, 28–29 November 1996*, ed. Yoshihisa Kohno, 1–9. Tokyo: Central Research Institute of the Electric Power Industry (CRIEPI).

Fujita, Shinichi, Yoichi Ichikawa, Robert K. Kawaratani, and Yutaka Tonooka. 1991. Preliminary Inventory of Sulfur Dioxide Emissions in East Asia. *Atmospheric Environment* 25(7): 1409–1411.

Fujita, Shinichi, and Robert K. Kawaratani. 1988. Wet Deposition of Sulfate in the Inland Sea Region of Japan. *Journal of Atmospheric Chemistry* 7: 59–72.

Fujita, Shinichi, Akira Takahashi, and Shō Nishinomiya. 1994. Wagakuni niokeru Sanseiu no Jittai: Kōsui no Kansokumō no Kōchiku (Acidic Precipitation in Japan: Design of the CRIEPI Monitoring Network for the Chemistry of Precipitation). *Kankyō Kagaku Kaishi (Journal of Japanese Society of Environmental Sciences)* 7(2): 107–120.

Fujita, Shinichi, and Nobuyuki Terada. 1985. Zensensei no Kōu ni Tomonau Ryūsanion Nōdo to Chinchakuryō no Hendōtokusei nitsuite (On the Temporal Variation of Concentration and Wet Deposition of Sulfate Associated with a Single Event of Frontal Depression). *Taiki Osen Gakkaishi (Journal of Japanese Society of Air Pollution)* 20(3): 188–197.

Fukawa, Satoru, and Katsuzō Kamiyama. 1994. *Tanaka Shōzō to Ashio Kōdoku Jiken o Aruku (Guidebook to Places Associated with Tanaka Shōzō and the Ashio Copper Mine Poisoning Incident).** Utsunomiya, Japan: Zuisōsha.

Fukui, Eiichiro, ed. 1977. *The Climate of Japan.* Tokyo: Kodansha.

Galloway, James N., Dianwu Zhao, Jiling Xiong, and Gene E. Likens. 1987. Acid Rain: China, United States, and a Remote Area. *Science* 236: 1559–1562.

Garrety, Karin. 1997. Social Worlds, Actor-Networks and Controversy: The Case of Cholesterol, Dietary Fat and Heart Disease. *Social Studies of Science* 27: 727–773.

Geochemical Laboratory, Meteorological Research Institute. 1978. *Geochemical Study of the Ocean and the Atmosphere: Yasuo Miyake, Seventieth Anniversary, Collected Papers (1939–1977).* Tokyo: Geochemical Laboratory, Meteorological Research Institute.

George, Timothy S. 2001. *Minamata: Pollution and the Struggle for Democracy in Postwar Japan.* Cambridge, MA: Harvard University Press.

Glacken, Clarence J. 1967. *Traces on the Rhodian Shore: Nature and Culture in Western Thought from Ancient Times to the End of the Eighteenth Century.* Berkeley: University of California Press.

Gorham, Eville. 1981. Scientific Understanding of Atmosphere-Biosphere Interactions: A Historical Overview (chapter 2). In *Atmosphere-Biosphere Interactions: Towards a Better Assessment of the Ecological Consequences of Fossil Fuel Combustion*, ed. Committee on the Atmosphere and the Biosphere, U.S. National Research Council, 9–21. Washington, DC: National Academy Press.

———. 1989. Scientific Understanding of Ecosystem Acidification: A Historical Review. *Ambio* 18(3): 150–154.

Gould, Roy. 1985. *Going Sour: Science and Politics of Acid Rain.* Boston: Birkhäuser.

Gresser, Julian, Koichiro Fujikura, and Akio Morishima. 1981. *Environmental Law in Japan*. Cambridge, MA: MIT Press.

Haas, Ernst B. 1990. *When Knowledge Is Power*. Berkeley: University of California Press.

Haas, Peter M. 1989. Do Regimes Matter? Epistemic Communities and Mediterranean Pollution Control. *International Organization* 43(3): 377–403.

———. 1990a. *Saving the Mediterranean: The Politics of International Environmental Cooperation*. New York: Columbia University Press.

———. 1990b. Obtaining International Environmental Protection through Epistemic Consensus. *Millennium* 19(3): 347–363.

———. 1992a. Knowledge, Power, and International Policy Coordination. *International Organization* 46(1).

———. 1992b. Introduction: Epistemic Communities and International Policy Coordination. *International Organization* 46(1): 1–35.

———. 1992c. Banning Chlorofluorocarbons: Epistemic Community Efforts to Protect Stratospheric Ozone. *International Organization* 46(1): 186–224.

———. 1993a. Protecting the Baltic and North Seas. In *Institutions for the Earth: Sources of Effective International Environmental Protection*, eds. Peter M. Haas, Robert O. Keohane, and Marc A. Levy, 133–181. Cambridge, MA: MIT Press.

———. 1993b. Epistemic Communities and the Dynamics of International Environmental Co-Operation. In *Regime Theory and International Relations*, ed. Volker Rittberger, 168–201. Oxford: Clarendon Press.

———. 1997. *Knowledge, Power, and International Policy Coordination*. Columbia: University of South Carolina Press.

Hacking, Ian. 1983. *Representing and Intervening*. Cambridge, UK: Cambridge University Press.

Hara, Hiroshi. 1993. Acid Deposition Chemistry in Japan. *Kōshū Eiseiin Kenkyū Hōkoku (Bulletin of the Institute for Public Health)* 42(3): 426–437.

Hara, Hiroshi, Eiichi Ito, Takunori Katou, Yoko Kitamura, Tetsuhito Komeiji, Mayumi Oohara, Toshiichi Okita, Kyoichi Sekiguchi, Keisuke Taguchi, Motonori Tamaki, Yoshio Yamanaka, and Kenichiro Yoshimura. 1990. Analysis of Two-Year Results of Acid Precipitation Survey within Japan. *Bulletin of the Chemical Society of Japan* 63: 2691–2697.

Hashimoto, Michio. 1989. History of Air Pollution Control in Japan. In *How to Conquer Air Pollution: A Japanese Experience*, ed. Hajime Nishimura, 1–93. Amsterdam: Elsevier.

Hatakeyama, Shiro, Kentaro Murano, Hiroshi Bandow, Hitoshi Mukai, and Hajime Akimoto. 1995a. High Concentration of SO_2 Observed Over the Sea of Japan. *Terrestrial, Atmospheric and Oceanic Sciences* 6(3): 403–408.

Hatakeyama, Shiro, Kentaro Murano, Hiroshi Bandow, Fumio Sakamaki, Masahiko Yamato, Shigeru Tanaka, and Hajime Akimoto. 1995b. The 1991

PEACAMPOT Aircraft Observation of Ozone, NO_x, and SO_2 over the East China Sea, the Yellow Sea, and the Sea of Japan. *Journal of Geophysical Research* 100(D11): 23,143–23,151.

Hirasawa, Kenzō. 1941. Miyako ni okeru Kōsui Bunseki Kekka nitsuite (On the Results of Chemical Analyses of Rain Water in Miyako). *Kishō Shūshi (Journal of the Meteorological Society of Japan)* 19(10): 395–400.

Hironaka, Wakako. 1993. Towards a Strengthened Regional Cooperation in Northeast Asia. In *The Second Northeast Asian Conference on Environmental Cooperation (15–17 September), Seoul, Korea*, 47–50. Tokyo: OECC (Overseas Environmental Cooperation Center).

Hitachi Kinzoku K.K. (Hitachi Metals Co.). 1994. *Daientotsu no Kiroku: Hitachi Dōzan Engai Taisaku Shi (Record of the Big Stack: A History of the Hitachi Copper Mine's Smoke Damage Countermeasures).** Tokyo: Hitachi Kinzoku K.K.

Hitachi Kōgyō K.K. (Hitachi Mining Co.). 1991. *Kuhara Fusanosuke Shōden (A Biography of Kuhara Fusanosuke).** Tokyo: Hitachi Kōgyō K.K.

Holzner, Burkart. 1972. *Reality Construction in Society*. Cambridge, MA: Schenkman Publishing Co.

Hong, Sungmin, Jean-Pierre Candelone, Clair C. Patterson, and Claude F. Boutron. 1996. History of Ancient Copper Smelting Pollution During Roman and Medieval Times Recorded in Greenland Ice. *Science* 272: 246–248.

Hordijk, Leen. 1995. Integrated Assessment Models as a Basis for Air Pollution Negotiations. *Water, Air and Soil Pollution* 85: 249–260.

Hoshino, Yoshiro. 1992. Japan's Post-Second World War Environmental Problems. In *Industrial Pollution in Japan*, ed. Jun Ui, 64–76. Tokyo: United Nations University Press.

Howells, Gwyneth P. 1995. *Acid Rain and Acid Waters*. New York: Ellis Horwood.

Huang, Meiyuan, Zifa Wang, Dongyang He, Huaying Xu, and Ling Zhou. 1995. Modeling Studies on Sulfur Deposition and Transport in East Asia. *Water, Air and Soil Pollution* 85: 1873–1878.

Huddle, Norie, and Michael Reich. 1987. *Island of Dreams: Environmental Crisis in Japan*. Rochester, VT: Schenkman Books.

Ichikawa, Yoichi, and Shinichi Fujita. 1993. *Higashi Ajia Chiiki niokeru Ryūsan Ion no Shissei Chinchaku no Kaiseki (An Analysis of Wet Deposition of Sulfate in East Asia), Report T92041*. Tokyo: CRIEPI (Central Research Institute of the Electric Power Industry).

———. 1994. *Wagakuni no Ryūsan Ion no Shissei Chinchakuryō ni oyobosu Higashi Ajia Kakkoku no Kiyo Hyōka (An Estimation of the Contribution of East Asian Countries to Wet Deposition of Sulfate in Japan), Report #T93012*. Tokyo: CRIEPI (Central Research Institute of the Electric Power Industry).

———. 1995. An Analysis of Wet Deposition of Sulfate Using a Trajectory Model for East Asia. *Water, Air, and Soil Pollution* 85(4): 1927–1932.

Ichikawa, Yoichi, Shinichi Fujita, and Yūkō Ikeda. 1994. Higashi Ajia Chiiki o taishō toshita Torajekutorii-gata Moderu niyoru Ryūsan Ion no Shissei Chinchaku no Kaiseki (An Analysis of Wet Deposition of Sulfate Using a Trajectory Model for East Asia). *Doboku Gakkai Ronbunshū (Journal of Hydraulic, Coastal and Environmental Engineering)* 497/II-28: 127–136.

Iijima, Nobuko, ed. 1979. *Pollution Japan: Historical Chronology*. Tokyo: Asahi Evening News.

Imura, Hidefumi. 1994. Japan's Environmental Balancing Act: Accomodating Sustained Development. *Asian Survey* 34(4): 355–368.

Interim Secretariat of EANET. 1998a. *Acid Deposition Monitoring Network in East Asia: The First Intergovernmental Meeting (19–20 March 1998), The First Meeting of the Working Group (5–7 November 1997), The Second Meeting of the Working Group (17–18 March 1998)*. Tokyo: Environment Agency of Japan.

———. (1998b) *Acid Deposition Monitoring Network in East Asia: The First Meeting of the Interim Scientific Advisory Group (12–14 October 1998), The Third Meeting of the Working Group (12–14 October 1998)*. Tokyo: Environment Agency of Japan.

International Association for Meteorology and Atmospheric Sciences. 1996. *Proceedings of the International Conference on Acid Deposition in East Asia, Taipei, Taiwan, 28–30 May 1996*. Taipei: International Association for Meteorology and Atmospheric Sciences.

Ishikawa, Yuriko, and Hiroshi Hara. 1997. Historical Change in Precipitation pH at Kobe, Japan: 1935–1961. *Atmospheric Environment* 31(15): 2367–2369.

Itsukakai, ed. 1926. *Furukawa Ichibē Den (Biography of Furukawa Ichibei).** Tokyo: Itsukakai.

Jackson, C. Ian. 1990. A Tenth Anniversary Review of the ECE Convention on Long-Range Transboundary Air Pollution. *International Environmental Affairs* 2(3): 217–226.

Jacob, Anthony T. 1991. *Acid Rain: The Chemistry of Acid Deposition from the Atmosphere*. Madison, WI: Institute for Chemical Education, University of Wisconsin–Madison.

Japan Times. 1994. Link between China, acid rain eyed. *Japan Times* (5 July), 3.

Jasanoff, Sheila S. 1987. Contested Boundaries in Policy-Relevant Science. *Social Studies of Science* 17(2): 195–230.

———. 1990. *The Fifth Branch*. Cambridge, MA: Harvard University Press.

Jasanoff, Sheila, G. E. Markle, J. C. Peterson, and T. Pinch, eds. 1996. *Handbook of Science and Technology Studies*. Thousand Oaks, CA: Sage.

JMA (Japan Meterological Agency). 1994. *Kinnen niokeru Sekai no Ijōkishō to Kikōhendō (Recent Unusual Global Weather Phenomena and Climate Changes).** Tokyo: JMA (Japan Meterological Agency).

Johnson, Chalmers. 1982. *MITI and the Japanese Miracle: The Growth of Industrial Policy 1925–1975*. Stanford, CA: Stanford University Press.

Jubak, Jim, and Marie D'Amico. 1993. Mighty MITI: Japan Wields Industrial Policy to Try to Become the World Leader in Environmental Technology. *Amicus* (summer): 38–43.

Kamioka, Namiko. 1987. *Hihon no Kōgaishi (History of Pollution in Japan).** Tokyo: Sekaishoin.

Kankyō Gijutsu (Environmental Conservation Engineering). 1985. Wagakuni Kakuchi niokeru Sanseiu Jittai Chōsa (Current Acid Rain Investigations in Various Districts in Japan).* *Kankyō Gijutsu (Environmental Conservation Engineering)* 14(2).

Kasuga, Hitoshi. 1989. Health Effects of Air Pollution. In *How to Conquer Air Pollution: A Japanese Experience*, ed. Hajime Nishimura, 95–113. Amsterdam: Elsevier.

Kato, Nobuo, and Hajime Akimoto. 1992. Anthropogenic Emissions of SO_2 and NO_x in Asia: Emission Inventories. *Atmospheric Environment* 26(16): 2997–3017.

Kato, Nobuo, Yoshiki Ogawa, Toshiya Koike, Tamotsu Sakamoto, and Susumu Sakamoto. 1991. *Analysis of the Structure of Energy Consumption and the Dynamics of Emissions of Atmospheric Species Related to the Global Environmental Change (SO_x, NO_x and CO_2) in Asia, NISTEP Report No. 21*. Tokyo: National Institute of Science and Technology Policy, 4th Policy-Oriented Research Group.

Kato, Saburo. 1989. System for Regulation. In *How to Conquer Air Pollution: A Japanese Experience*, ed. Hajime Nishimura, 197–238. Amsterdam: Elsevier.

Katō, Takeo. 1935. Ko no Ha ni Tsuku Ojin no Kagaku Seibun Kekka nitsuite—Dai ichi hō (Chemical Analysis of Soot Attached to Leaves—Part 1).* *Umi to Sora (Sea and Sky)* 15(4): 119–126.

———. 1939. Ko no Ha ni Tsuku Ojin no Kagaku Seibun Kekka nitsuite—Dai ni hō (Chemical Analysis of Soot Attached to Leaves—Part 2).* *Umi to Sora (Sea and Sky)* 19(10): 303–311.

Katoh, T., T. Konno, I. Koyama, H. Tsuruta, and H. Makino. 1990. Acidic Precipitation in Japan. In *Acidic Precipitation: Volume 5 (International Overview and Assessment)*, eds. A. H. M. Bresser and W. Salomons, 41–105. New York: Springer-Verlag.

Katz, M., ed. 1939. *Effect of Sulfur Dioxide on Vegetation (Report #815)*. Ottawa: National Research Council of Canada.

Kellner, Osker. 1887. Tokyo Usuichū no Ammonia oyobi Shōsan no Ryō (Ammonia and Nitrate Content of Rainwater in Tokyo).* *Tokyo Kagaku Kaishi (Bulletin of the Chemical Society of Tokyo)** 8: 161–180.

———. 1894. Usuichū no Ammonia oyobi Shōsan no Ryō (Ammonia and Nitrate Content of Rainwater).* *Nōka Daigaku Gakujitsu Jiken Ihō (Technical Report of the Agricultural University of Tokyo)** 1: 28–42.

Keohane, Robert O. 1989. Neoliberal Institutionalism: A Perspective on World Politics. In *Institutions and State Power: Essays in International Relations Theory*, ed. Robert O. Keohane, 1–20. Boulder: Westview Press.

———. 1993. The Analysis of International Regimes. In *Regime Theory and International Relations*, ed. Volker Rittberger, 23–48. Oxford: Clarendon Press.

Kishimoto, Kōichi. 1988. *Politics in Modern Japan: Development and Organization*. Tokyo: Japan Echo Inc.

Kitamura, Moritsugu, Takunori Katou, Kyoichi Sekiguchi, Keisuke Taguchi, Motonori Tamaki, Mayumi Oohara, Atsuko Mori, Kentaro Murano, Shinji Wakamatsu, Yoshio Yamanaka, Toshiichi Okita, and Hiroshi Hara. 1991. Wagakuni no Sanseiu no pH to sono Hindo Bunpu Patan (pH and its Frequency Distribution Patterns of Acid Precipitation in Japan). *Nippon Kagaku Kaishi (Journal of the Chemical Society of Japan)* 6: 913–919.

Kodansha. 1983. *Kodansha Encyclopedia of Japan*. Tokyo: Kodansha.

Kohno, Yoshishisa. 1987. *Effect of Simulated Acid Rain on the Growth and Yield of Soybean Plants Grown in Pots, Report U87015*. Tokyo: CRIEPI (Central Research Institute of the Electric Power Industry).

Kohno, Yoshishisa, and Takuya Fujiwara. 1981. *Effect of Acid Rain on the Growth and Yield of Radish and Kidney Bean, Report E481008*. Tokyo: CRIEPI (Central Research Institute of the Electric Power Industry).

———. 1982. *Cryo-Scanning Electron Microscopic Observations and Concentrations of Inorganic Elements in Radish and Kidney Bean Exposed to Acid Rain, Report E481014*. Tokyo: CRIEPI (Central Research Institute of the Electric Power Industry).

Kohno, Yoshihisa, ed. 1997. *Proceedings of CRIEPI International Seminar on Transport and Effects of Acidic Substances, 28–29 November 1996, Tokyo, Japan*. Tokyo: CRIEPI (Central Research Institute of the Electric Power Industry).

Kohno, Yoshishisa, and Takuya Kobayashi. 1989a. Effect of Simulated Acid Rain on the Growth of Soybean. *Water, Air and Soil Pollution* 43: 11–19.

———. 1989b. Effect of Simulated Acid Rain on the Yield of Soybean. *Water, Air and Soil Pollution* 45: 173–181.

Krasner, Stephen D. 1983. Structural Causes and Regime Consequences: Regimes as Intervening Variables. In *International Regimes*, ed. Stephen D. Krasner, 1–22. Ithaca, NY: Cornell University Press.

Kuhn, Thomas S. 1962. *The Structure of Scientific Revolutions*. Chicago: University of Chicago Press.

Kurabayashi, Yōko. 1994. Koshōgaku no Sendachi (Limnology's Pioneers).* *Ikiru* 11(1): 30–33.

Kurashige, Eijirō. 1934. Ame no Seibun (Composition of Rain).* *Tenki to Kikō (Weather and Climate)** 1(9): 390–393.

Kurashige, Eijirō, and Gōsuke Kagei. 1935. Tokyo-shinai oyobi Kōgai niokeru Kōsui no Ondoku nitsuite (Dai 1 Hō) (Chemical Investigation on the Precipitations at Tokio and its Suburbs). *Kishō Shūshi (Journal of the Meteorological Society of Japan)* 13(5): 211–216.

Kurita, Hedemi, Jyunichi Hori, Yoshio Hamada, and Hiromasa Ueda. 1993. Chūbu Sangaku Chiiki Kasen Jōryūiki niokeru Kasen, Koshō pH no Keinenteki Teika to Sanseiu no Kankei nitsuite (Decrease of pH of River and Lake Water in Mountainous Region in Central Japan and Its Relation to Acid Rain). *Taiki Osen Gakkaishi (Journal of Japanese Society of Air Pollution)* 28(5): 308–315.

Kurokawa, Michiko, Toshiaki Asō, Haruo Mimura, Keiji Aihara, Mikihiro Kaneko, Shinichi Nishiyama, Kōji Himi, and Saburō Sugano. 1975. Kōuchū no Horumuarudehido Nōdo to Me ni taisuru Shigekisei ni tsuite (Concentration of Formaldehyde in Rain Water and the Eye Irritation due to Aldehyde Group). *Taiki Osen Kenkyū (Journal of the Japan Society of Air Pollution)* 10(4): 628.

Lakatos, Imre. 1970. Falsification and the Methodology of Scientific Research Programmes. In *Criticism and the Growth of Knowledge*, eds. Imre Lakatos and Alan Musgrave, 91–195. London: Cambridge University Press.

Lapp, Ralph E. 1958. *The Voyage of the Lucky Dragon*. New York: Harper & Brothers Publishers.

Latour, Bruno. 1987. *Science in Action*. Cambridge, MA: Harvard University Press.

———. 1988. *The Pasteurization of France*. Cambridge, MA: Harvard University Press.

Latour, Bruno, and Steve Woolgar. 1986. *Laboratory Life: The Social Construction of Scientific Facts*. Princeton: Princeton University Press.

Lehman, Jean-Pierre. 1978. *The Image of Japan: From Feudal Isolation to World Power, 1850–1905*. London: Allen & Unwin.

Lélé, Sharachchandra M. 1991. Sustainable Development: A Critical Review. *World Development* 19(6): 607–621.

Levy, Marc A. 1993. European Acid Rain: The Power of Tote-Board Diplomacy. In *Institutions for the Earth: Sources of Effective International Environmental Protection*, eds. Peter M. Haas, Robert O. Keohane, and Marc A. Levy, 75–132. Cambridge, MA: MIT Press.

Likens, Gene E., F. Herbert Bormann, and Noye M. Johnson. 1972. Acid Rain. *Environment* 14(2): 33–40.

Likens, Gene E., and Thomas J. Butler. 1981. Recent Acidification of Precipitation in North America. *Atmospheric Environment* 15(7): 1103–1109.

Liu, Chung-Ming, Fu-Tien Jeng, Shaw C. Liu, and Chea-Yuan Young. 1993. A Study of Acid Deposition in Taiwan. A paper presented at the International Workshop on the Harmonization of Monitoring Techniques of Acid Deposition and Methodology of Emission Inventories of SO_2 and NO_x in East Asia, 27–29 January, Tsukuba, Japan.

Lohmann, Larry. 1990. Whose Common Future? *The Ecologist* 20(3): 82–84.

Longhurst, James W. S. 1990. Acid Deposition. In *World Guide to Environmental Issues and Organizations*, ed. Peter Brackley, 3–23. Harlow: Longman Current Affairs.

Maeda, Yasuaki. 1995. Higashi Ajia Chiiki o Taishō toshita Sansei Taiki Osen Busshitsu no Bunkazai oyobi Zairyo e no Eikyō Chōsa—Dai 3 Hō (Effects of Acidic Air Pollutants on Cultural Properties and Materials in the East Asia Region—Report #3).* In *Taiki Kankyō Gakkai Nenkai—Dai 36 Kai (Proceedings of the 36th Annual Meeting of the Atmospheric Environment Society of Japan), Tokyo*, 199–200. Tokyo: Taiki Kankyō Gakkai (Atmospheric Environment Society of Japan).

Maruyama, Hiroshi, Motonori Tamaki, Mitsuru Shoga, and Takatoshi Hiraki. 1993. Sanseiu Kenkyū no Bunkateki Isan (1): Sore wa Shōwa 10 Nen, Kobe kara Hajimatta (The Cultural Heritage of Acid Rain Research: It Started in Kobe in 1935).* *Kankyō Gijutsu (Environmental Conservation Engineering)* 22(12): 732–735.

Matsudaira, Yasuo. 1933. Kobe no Yuki nitsuite (On the Snow in Kobe).* *Umi to Sora (Sea and Sky)* 13(5): 111–113.

———. 1937. Taifū Kennai ni Haitta ori no Kobe niokeru Kōsui no Kagaku Seibun (Chemical Composition of Typhoon Precipitation in Kobe).* *Umi to Sora (Sea and Sky)* 17(11): 420–425.

———. 1938. Kobe de Kansokushita Kōsa nitsuite (On the Measurement of Yellow Sands in Kobe).* *Umi to Sora (Sea and Sky)* 18(7): 231–235.

Matsudaira, Yasuo, and Takeo Katō. 1933. Kobe niokeru Natsu no Usuichū no Nisan Kagaku Seibun nitsuite (On the Chemical Composition of Summer Rain in Kobe).* *Umi to Sora (Sea and Sky)* 13(10): 265–269.

Matsui, Hideo. 1939. Tokyo Marunouchi Fukin no Kūkichū Raizatsubutsu nitsuite (Atmospheric Impurities in the Central Part of Tokyo). *Kishō Shūshi (Journal of the Meteorological Society of Japan)* 17(9): 367–372.

———. 1941. Tokyo Marunouchi Fukin no Kūkichū Raizatsubutsu nitsuite—Dai San Hō (Atmospheric Impurities in the Central Part of Tokyo—Part III). *Kishō Shūshi (Journal of the Meteorological Society of Japan)* 19(3): 72–76.

———. 1942a. Maebashi-shi Fukin no Kūkichū Raizatsubutsu nitsuite (Atmospheric Impurities in the City of Maebasi). *Kishō Shūshi (Journal of the Meteorological Society of Japan)* 20(3): 94–97.

———. 1942b. Usui no Kagaku (Dai Ni Hō) (The Chemistry of Rain Water. Part II). *Kishō Shūshi (Journal of the Meteorological Society of Japan)* 20(8): 291–296.

McCormick, John. 1989. *Acid Earth: The Global Threat of Acid Pollution (2nd Edition)*. London: Earthscan Publications, Ltd.

McKean, Margaret A. 1981. *Environmental Protest and Citizen Politics in Japan*. Berkeley: University of California Press.

Meteorological Society of Japan, ed. 1987. *Sanseiu (Acid Rain),* Report 158.* Tokyo: Nihon Kishō Gakkai (Meteorological Society of Japan).

Michio, Nagai, and Miguel Urrutia, eds. 1985. *Meiji Ishin: Restoration and Revolution.* Tokyo: United Nations University.

Miller, Alan S., and Curtis Moore. 1991. *Japan and the Global Environment.* College Park, MD: Center for Global Change, University of Maryland.

Miller, Clark A., and Paul N. Edwards, eds. 2001. *Changing the Atmosphere: Expert Knowledge and Environmental Governance.* Cambridge, MA: MIT Press.

Miller, David, ed. 1985. *Popper Selections.* Princeton: Princeton University Press.

Miyake, Yasuo. 1937. Taikichū Raizatsubutsu nitsuite (Atmospheric Impurities in Central Tokyo). *Kishō Shūshi (Journal of the Meteorological Society of Japan)* 15(12): 529–531.

———. 1938. Tokyo Marunouchi Fukin no Kūkichū Raizatsubutsu nitsuite (Atmospheric Impurities in Central Tokyo). *Kishō Shūshi (Journal of the Meteorological Society of Japan)* 16(12): 461–463.

———. 1939. Usui no Kagaku (The Chemistry of Rain Water). *Kishō Shūshi (Journal of the Meteorological Society of Japan)* 17(1): 20–37.

———. 1950. Kōsui no Kagaku Seibun nitsuite (On the Chemical Composition of Precipitation).* *Suidō Kyōkai Zasshi (Journal of the Waterworks and Sewerage Association)* 188: 20–24.

———. 1954. *Chikyū Kagaku (Earth Chemistry).** Tokyo: Asakura Shōten.

———. 1957. *Kōsui no Kagaku (Precipitation Chemistry).** Tokyo: Chijin Shokan.

———. 1969. Suishitsu nitsuite no Ikutsuka no Dammen (The Many Faces of Water).* *Suidō Kyōkai Zasshi (Journal of the Japan Water Works Association)* 412: 16–21.

MOE (Ministry of Environment of the Republic of Korea). 1995. *Environmental Protection in Korea, 1995.* Seoul: Ministry of Environment of the Republic of Korea.

Mohnen, Volker A. 1988. The Challenge of Acid Rain. *Scientific American* 259(Aug): 14–22.

Moist Air Pollution Investigation Committee. 1975. *Shissei Taiki Osen (Sansei Kōu) nitsuite (About Moist Air Pollution (Acid Precipitation)).** Tokyo: Environment Agency Air Protection Bureau.

———. 1976. *Shōwa 50 Nendo Shissei Taiki Osen Chōsa Kekka Hōkokusho (Report on the Results of the 1975 Fiscal Year Moist Air Pollution Investigation).** Tokyo: Environment Agency Air Protection Bureau.

———. 1977. *Shōwa 51 Nendo Shissei Taiki Osen Chōsa Kekka Hōkokusho (Report on the Results of the 1976 Fiscal Year Moist Air Pollution Investigation).** Tokyo: Environment Agency Air Protection Bureau.

———. 1978. *Shōwa 52 Nendo Shissei Taiki Osen Chōsa Kekka Hōkokusho (Report on the Results of the 1977 Fiscal Year Moist Air Pollution Investigation).** Tokyo: Environment Agency Air Protection Bureau.

———. 1979. *Shōwa 53 Nendo Shissei Taiki Osen Chōsa Kekka Hōkokusho (Report on the Results of the 1978 Fiscal Year Moist Air Pollution Investigation).** Tokyo: Environment Agency Air Protection Bureau.

Moore, Curtis, and Alan S. Miller. 1994. *Green Gold: Japan, Germany, the United States, and the Race for Environmental Technology.* Boston: Beacon Press.

Mori, Atsuko, Mayumi Ohara, Shinji Wakamatsu, Kentaro Murano, Keisuke Taguchi, Kyoichi Sekiguchi, Motonori Tamaki, Hironori Kato, Moritsugu Kitamura, Toshiichi Okita, Yoshio Yamanaka, and Hiroshi Hara. 1991. Sansei Chinchakubutsu niokeru Shōsan Ion to Ryūsan Ion no Tōryōhi ni kansuru Kōsatsu (Studies on Equivalent Ratio of Nitrate to Sulfate in Acid Precipitation). *Nippon Kagaku Kaishi (Journal of the Chemical Society of Japan)* 6: 920–929.

Morikawa, Yasushi. 1989. Nihon no Shinrin Suitai to Sanseiu Mondai: Kantō Heiya no Sugi no Suitai o Rei toshite (The Acid Rain Problem and Japanese Forests: The Case of *Sugi* Decline in the Kanto Plain).* *Sanrin (Japanese Journal of Forestry)* 6: 10–17.

Morikawa, Y., Y. Maruyama, N. Tanaka, and T. Inoue. 1990. Forest Declines in Japan: Mature *Cryptomeria japonica* Declines in the Kanto Plains. In *Proceedings of the XIX World Congress of the IUFRO, Montreal*, 397–405. Vienna: International Union of Forestry Research Organizations (IUFRO).

Motokawa, Tatsuo. 1989. Sushi Science and Hamburger Science. *Perspectives in Biology and Medicine* 32(4): 489–504.

Munton, Don, Marvin Soroos, Elena Nikitina, and Marc A. Levy. 1999. Acid Rain in Europe and North America. In *The Effectiveness of International Environmental Regimes*, ed. Oran R. Young, 155–247. Cambridge, MA: MIT Press.

Murano, Kentaro. 1993. *Sanseiu to Sansei-Giri (Acid Rain and Acid Fog).** Tokyo: Shōkabō.

Murano, Kentaro, Yutaka Tonooka, Katsunori Suzuki, and Eisaku Toda. 1999. Emission Inventory Research Activities in Japan. A paper presented at the European Union Task Force on Emission Inventories, June 1999.

Murdo, Pat. 1990. *Japan's Environmental Policies: The International Dimension, Report 10A.* Washington, DC: Japan Economic Institute.

———. 1992. *Environmental Developments Offer Opportunities for Japan, Report 1A.* Washington, DC: Japan Economic Institute.

Myers, Frederick S. 1992. Japan Bids for Global Leadership in Clean Industry. *Science* 256: 1144–1145.

Nagpal, Tanvi. 1995. Voices from the Developing World: Progress Toward Sustainable Development. *Environment* 37(8): 10–15 and 30–35.

Nakai, Nobuyuki, and Ushio Takeuchi. 1974. Ame no Kagaku to Taiki Osen (Rainwater Chemistry and Air Pollution).* *Kagaku (Chemistry)* 29(6): 418–426.

Nakane, Chie, and Shinzaburō Oishi, eds. 1991. *Tokugawa Japan: The Social and Economic Antecedents of Modern Japan.* Tokyo: University of Tokyo Press.

NAPAP (National Acid Precipitation Assessment Program). 1990. *(1) NAPAP 1990 Integrated Assessment Report; (2) Acidic Deposition: State of Science and Technology Summary Report; (3) Acidic Deposition: State of Science and Technology Volumes I–IV—Volume I: Emissions, Atmospheric Processes and Deposition; Volume II: Aquatic Processes and Effects; Volume III: Terrestrial, Materials, Health and Visibility Effects; Volume IV: Control Technologies, Future Emissions, and Effects Valuation.* Washington, DC: NAPAP.

NIES (National Institute for Environmental Studies). 1996. *Proceedings of the International Symposium on Acidic Deposition and its Impacts, 10–12 December 1996, Tsukuba, Japan.* Tsukuba, Japan: NIES.

Nikei Keizai Shimbun. 1992. Fuyu no Tairiku kara Nishi Nihon ni Hirai, Haba 500 kiro, Sanseiu ni Eikyō (500-Kilometer-Wide Polluted Winter Air Mass from the Continent Hits Western Japan).* *Nikei Keizai Shimbun* (8 October), 19.

Nimura, Kazuo. 1997. *The Ashio Riot of 1907: A Social History of Mining in Japan.* Durham, NC: Duke University Press.

Nishi, Teizō. 1971. Usui no pH kara Mita Taiki Osen (Looking at Air Pollution from the Viewpoint of pH).* *Taiki Osen Nyūsu (Air Pollution News Report)* 64(4): 4–5.

Nishimura, Hajime, ed. 1989. *How to Conquer Air Pollution: A Japanese Experience.* Amsterdam: Elsevier.

Noguchi, Isamu. 1990. Sanseiu no Nōsakumotsu oyobi Shinrinboku e no Eikyō (Effects of Acid Precipitation on Agricultural Crops and Forest Trees). *Taiki Osen Gakkaishi (Journal of Japanese Society of Air Pollution)* 25(5): 295–312.

Notehelfer, F. G. 1975. Japan's First Pollution Incident. *Journal of Japanese Studies* 1(2): 351–383.

Odén, Svante. 1967. Nederbördens och Luftens Försurning-Dess Orsaker, Förlopp och Verkan I Olika Miljöer (The Acidification of Air and Precipitation and Its Consequences on the Natural Environment), Bulletin 1. Stockholm: Statens Naturvetenskapliga Forskningsråd (Swedish Natural Science Research Council), Ekologikommittén (Ecology Committee).

———. 1968. The Acidification of Air and Precipitation and Its Consequences on the Natural Environment, Tr-1172. Arlington, VA: Translation Consultants, Ltd.

OECC (Overseas Environmental Cooperation Center). 1992. *Northeast Asian Conference on Environmental Cooperation (13–16 October), Niigata, Japan.* Tokyo: OECC.

———. 1993. *Second Northeast Asian Conference on Environmental Cooperation (15–17 September), Seoul, Korea.* Tokyo: OECC.

OECD (Organization for Economic Co-operation and Development). 1977. *The OECD Programme on Long Range Transport of Air Pollutants: Summary Report.* Paris: OECD.

———. 1994. *OECD Environmental Performance Reviews: Japan.* Paris: OECD.

Ogura, Yutaka. 1940. Utsunomiya ni okeru Usui Seibun to Kishō Yōso no Kankei (The Chemical Composition of the Rain Water in Utsunomiya during 1937 and 1938). *Kishō Shūshi (Journal of the Meteorological Society of Japan)* 18(3): 107–111.

Ohta, Sachio, Toshiichi Okita, and Chiaki Kato. 1981. A Numerical Model of Acidification of Cloud Water. *Kishō Shūshi (Journal of the Meteorological Society of Japan)* 59(6): 892–901.

Oiwa, Keibo, Masato Ogata, and Karen Colligan-Taylor. 2001. *Rowing the Eternal Sea: The Story of a Minamata Fisherman.* Lanham, MD: Rowman and Littlefield.

Okamoto, Shinichi, and Noritaka Katatani. 1988. Taiki Osen no Chōkyori Yusō Moderu (Long-Range Transport Model for Air Pollution). *Tenki (Weather)** 35(8): 461–478.

Okita, Toshiichi. 1965. Some Chemical and Meteorological Measurements of Air Pollution in Asahikawa. *International Journal of Air and Water Pollution* 9: 323–332.

———. 1967. Jōon niokeru Aryūsan Gasu no Kyūchaku oyobi Sanka (Adsorption and Oxidation of Sulfur Dioxide at Ordinary Temperature). *Kōshū Eiseiin Kenkyū Hōkoku (Bulletin of the Institute of Public Health)* 16(2): 52–58.

———. 1968. Concentration of Sulfate and Other Inorganic Materials in Fog and Cloud Water and in Aerosol. *Kishō Shūshi (Journal of the Meteorological Society of Japan)* 46(2): 120–127.

———. 1971a. Barringer Sokan Supekutoromeeta niyoru Kawasaki-shi Kōgyōchitai kara no SO_2, NO_2 no Chiiki Hasseiryō oyobi Mihara-san kara no SO_2 no Hasseiryō no Sokutei (Measurements of areal emission of SO_2 and NO_2 at Kawasaki industrial district and of SO_2 emission from Mt. Mihara using a Barringer correlation spectrometer). *Kōshū Eiseiin Kenkyū Hōkoku (Bulletin of the Institute of Public Health)* 20(1): 47–53.

———. 1971b. Barringer Sokan Supekutoromeetaa niyoru Aryūsan Gasu oyobi Nisankaschiso Hasseiryō no Sokutei (Measurement of SO_2 and NO_2 Emissions using a Barringer Correlation Spectrometer).* Tokyo: Jasco (Japan Spectroscopic Company, Ltd.) of Tokyo.

———. 1972. Calculation of Rate of Absorption of Sulfur Dioxide by Rain- and Cloud-Droplets. *Kōshū Eiseiin Kenkyū Hōkoku (Bulletin of the Institute of Public Health)* 21(1): 9–13.

———. 1975. Kazan Gasu no Rimootosenshingu—Kazan kara Hoshutsusareru SO_2 no Sokutei (Remote Sensing Measurements of Mass Flow of Sulfur Dioxide Gas from Volcanoes). *Kazan (Volcanoes)** 19(3): 151–157.

———. 1983. Acid Precipitation and Related Phenomena in Japan. *Water Quality Bulletin (Environment Canada)* 8(2): 101–108.

———. 1995. Taiki to Torikunde 45 Nen—Nihon ni okeru Taiki Osen Earozoru, Sanseiu no Kenkyū no Yoake (45 Years Personal History of Air Research—Dawn of Studies of Air Pollution, Aerosol and Acid Deposition in Japan). *Aerozoru Kenkyū (Aerosol Research)** 10(4): 304–310.

Okita, Toshiichi, and Tetsuhito Komeiji. 1983. Showa 37 Nendo to Genjiten de no Tokyo-to niokeru Sansei, Arukarisei Busshitsu no Kōkaryō no Hikaku (A Comparison between the 1962/63 and 1980/81 Composition of Acidic and Alkaline Substances in Rainwater in Tokyo).* In *Taiki Osen Gakkai Nenkai—Dai 24 Kai (Proceedings of the 24th Annual Conference of the Japan Society of Air Pollution), Tokyo*, 550. Tokyo: Taiki Kankyō Gakkai (Atmospheric Environment Society of Japan).

Okita, Toshiichi, and Seishi Konno. 1964. Tokyo-to niokeru Kōkajinchū no Yōkaisei Muki Seibun no Bunseki (Chemical Analysis of Inorganic Water Soluble Components in Deposited Dust in Tokyo). *Kōshū Eiseiin Kenkyū Hōkoku (Bulletin of the Institute for Public Health)* 13(2): 121–125.

Overrein, L. N., H. M. Seip, and A. Tollan. 1980. *Acid Precipitation—Effects on Forest and Fish. Final Report of the SNSF Project 1972–1980.* Oslo: SNSF Project.

Overrein, Lars N. 1976. A Presentation of the Norwegian Project "Acid Precipitation—Effects on Forest and Fish." *Water, Air, and Soil Pollution* 6(2/3/4): 167–172.

Pearce, Fred. 1986. The Strange Death of Europe's Trees. *New Scientist* (4 December): 41–45.

Photochemical Formation of Secondary Pollutants Investigation Committee—Moist Air Pollutants Subcommittee. 1981. *Shissei Taiki Osen Chōsa Kekka Sōgō Hōkokusho (Sōkatsuhen) (Final Report on the Results of the Moist Air Pollution Investigation [Summary Edition]).** Tokyo: Environment Agency Air Protection Bureau.

Ponting, Clive. 1993. *A Green History of the World: The Environment and the Collapse of Great Civilizations.* New York: Penguin Books.

Pringle, Laurence P. 1988. *Rain of Troubles: The Science and Politics of Acid Rain.* New York: Macmillan.

Qi, Ling, Jiming Hao, and Mingming Lu. 1995. SO_2 Emission Scenarios of Eastern China. *Water, Air and Soil Pollution* 85: 1873–1878.

Quan, Hao. 1991. Chūgoku niokeru Sanseiu no Genjō to korekara no Kadai (Present Status of the Acid Rain Problem in China and Tasks for the Period Ahead). *Taiki Osen Gakkaishi (Journal of Japanese Air Pollution Society)* 26(5): 283–291.

———. 1993. Introduction to the Acid Precipitation Issue and its Control Measures in China. In *The Expert Meeting on Acid Precipitation Monitoring Network in East Asia (26–28 October), Toyama, Japan*, 95–105. Tokyo: Environment Agency of Japan.

Regens, James L., and Robert W. Rycroft. 1988. *The Acid Rain Controversy.* Pittsburgh, PA: University of Pittsburgh Press.

Reischauer, Edwin O. 1981. *The Japanese.* Cambridge, MA: Harvard University Press.

———. 1990. *Japan: The Story of a Nation, 3rd Edition.* Tokyo: Charles E. Tuttle Co.

RMCC (Federal/Provincial Research and Monitoring Coordinating Committee for the National Acid Rain Research Program). 1990. *1990 Canadian Long-Range Transport of Air Pollutants and Acid Deposition Assessment Report.* Ottawa: Environment Canada.

Rodhe, Henning. 1989. Acidification in a Global Perspective. *Ambio* 18(3): 155–160.

Rodhe, Henning, and Rafael Herrera, eds. 1988. *Acidification in Tropical Countries. SCOPE Report 36.* Chichester, England: Wiley.

Rosenstone, Robert A. 1988. *Mirror in the Shrine: American Encounters with Meiji Japan.* Cambridge, MA: Harvard University Press.

Ruggie, John G. 1975. International Responses to Technology: Concepts and Trends. *International Organization* 29(3): 557–583.

Ryaboshapko, A. G., V. V. Sukhenko, and S. G. Paramonov. 1994. Assessment of Wet Sulphur Deposition over the Former USSR. *Tellus* 46B(3): 205–219.

Sand, Peter. 1987. Air Pollution in Europe: International Policy Responses. *Environment* 29(10): 16–20, 28–29.

Saruhashi, Katsuko, and Teruko Kanazawa. 1978. Kōsui no pH (Precipitation pH).* *Tenki (Weather)** 25(11): 2–4.

Satake, Kenichi, ed. 1999. *Sansei Kankyō no Seitaigaku (Ecology of Acidic Environments).** Tokyo: Aichi Shuppan.

Satake, Kenichi, Akira Oyagi, and Yasuko Iwao. 1995. Natural Acidification of Lakes and Rivers in Japan: The Ecosystem of Lake Usoriko (pH 3.4–3.8). *Water, Air and Soil Pollution* 85: 511–516.

Satake, Kenichi, and Yatsuka Saijo. 1974. Carbon Dioxide Content and Metabolic Activity of Microorganisms in Some Acid Lakes in Japan. *Limnology and Oceanography* 19(2): 331–338.

Sato, Junji, Hidetaka Sasaki, and Takehiko Satomura. 1999. Transport of Sulfur Oxides over the East Asian Region by the Off-line Coupled Meteorological and Transport Model. *Papers in Meteorology and Geophysics* 50(3): 97–111.

Schmandt, Jurgen, Judith Clarkson, and Hilliard Roderick. 1988. *Acid Rain and Friendly Neighbors: The Policy Dispute Between Canada and the United States (Revised Edition).* Durham, NC: Duke University Press.

Schreurs, Miranda A. 1994. Policy Laggard or Policy Leader? Global Environmental Policy-Making Under the Liberal Democratic Party. *Journal of Pacific Asia* 2: 3–33.

Schreurs, Miranda A., Patricia Welch, and Akiko Kōda. 1995. Japan: Elite Newspaper Reporting on the Acid Rain Issue from 1972 to 1992. In *The Press and*

Global Environmental Change: An International Comparison of Elite Newspaper Reporting on the Acid Rain Issue from 1972 to 1992 (Environment and Natural Resources Program Working Paper No. E-95-06), eds. William C. Clark and Nancy M. Dickson, pp. G-1–G-36. Cambridge, MA: Center for Science and International Affairs, John F. Kennedy School of Government, Harvard University.

Sebenius, James K. 1992. Challenging Conventional Explanations of International Cooperation: Negotiation Analysis and the Case of Epistemic Communities. *International Organization* 46(1): 323–365.

Seinfeld, John H. 1986. *Atmospheric Chemistry and Physics of Air Pollution.* New York: Wiley.

Seip, Hans M., Dianwu Zhao, Jiling Xiong, Dawei Zhao, Thorjorn Larssen, Bohan Liao, and Rolf D. Vogt. 1995. Acidic Deposition and its Effects in Southwestern China. *Water, Air, and Soil Pollution* 85: 2301–2306.

Sekiguchi, K., Y. Hara, and A. Ujiiye. 1986. Dieback of Cryptomeria Japonica and Distribution of Acid Deposition and Oxidant in Kanto District of Japan. *Environmental Technology Letters* 7(5): 263–268.

Sekiguchi, Koichi, Kazuo Kano, and Atsuo Ujiiye. 1983. Maebashi-shi ni Futta pH 2.86 no Ame nitsuite (Acid Rain (pH 2.86) in Maebashi). *Taiki Osen Gakkaishi (Journal of Japanese Society of Air Pollution)* 18(1): 1–7.

Seliga, Thomas A., and Leon S. Dochinger. 1976. First International Symposium on Acid Precipitation and the Forest Ecosystem. *Water, Air and Soil Pollution* 6(2/3/4): 135.

Shigeru, Nakayama, David L. Swain, and Yagi Eri, eds. 1974. *Science and Society in Modern Japan: Selected Historical Sources.* Cambridge, MA: MIT Press.

Shimizu, Mitsuo. 1936. Hamamatsu niokeru Usuichū no Nisan Kagaku Seibun to Kaze no Kankei nitsuite (On the Chemical Analysis of Rainwater in Hamamatsu and its Relation to Wind).* *Tenki to Kikō (Weather and Climate)** 3: 256–260.

Shindo, J., A. K. Bregt, and T. Hakamata. 1995. Evaluation of Estimation Methods and Base Data Uncertainties for Critical Loads of Acid Deposition in Japan. *Water, Air and Soil Pollution* 85: 2571–2576.

Shoji, Kichiro, and Masuro Sugai. 1992. The Ashio Copper Mine Pollution Case: The Origins of Environmental Destruction. In *Industrial Pollution in Japan*, ed. Jun Ui, 18–63. Tokyo: United Nations University Press.

Simmons, I. G. 1996. *Changing the Face of the Earth, 2nd Edition.* Oxford: Blackwell.

Soroos, Marvin S. 1997. *The Endangered Atmosphere: Preserving a Global Commons.* Columbia, SC: University of South Carolina Press.

Star, S. Leigh, and James R. Griesemer. 1989. Institutional Ecology, "Translations," and Boundary Objects: Amateurs and Profesionals in Berkeley's Museum of Vertebrate Zoology, 1907–1939. *Social Studies of Science* 19: 387–420.

State Council of China. 1992. *National Report of the People's Republic of China on Environment and Development (prepared for the UNCED "Earth Summit").* Beijing: China Environmental Science Press.

Stoiber, Richard E., and Anders Jepsen. 1973. Sulfur Dioxide Contributions to the Atmosphere by Volcanoes. *Science* 182: 577–578.

Stone, Alan. 1975. The Japanese Muckrakers. *Journal of Japanese Studies* 1(2): 385–407.

Streets, David G., Gregory R. Carmichael, Markus Amann, and Richard L. Arndt. 1999. Energy Consumption and Acid Deposition in Northeast Asia. *Ambio* 28(2): 135–143.

Strong, Kenneth. 1995. *Ox against the Storm: A Biography of Tanaka Shōzo (originally published in 1977).* Sandgate, UK: Japan Library.

Sugawara, Ken. 1948. Kōsui no Kagaku (Precipitation Chemistry).* *Kagaku (Science)** 18(11): 485–492.

Sugimoto, Masayoshi, and David L. Swain. 1978. *Science and Culture in Traditional Japan: A.D. 600–1854.* Cambridge, MA: MIT Press.

Summers, P. W., and D. M. Whelpdale. 1976. Acid Precipitation in Canada. *Water, Air, and Soil Pollution* 6(2/3/4): 447–455.

Swedish NGO Secretariat on Acid Rain. 1995. Environmental Factsheet No. 6—Critical Loads. *Acid News* 2 (April).

Takahashi, Keiji, Makoto Nashimoto, and Hiromasa Ueda. 1991. Kansai, Setouchi Chihō niokeru Sugi Suitai to Okishidanto Shisū, Kōuryō to no Kankei (Relationships among Oxidant Index, Precipitation and Decline of Japanese Cedar (*Cryptomeria japonica* D. Don) Trees in the Kansai-Setouchi District). *Kankyō Kagaku Kaishi (Journal of Environmental Science)* 4(1): 51–57.

Takahashi, Keiji, S. Okitsu, and Hiromasa Ueda. 1986. Kantō Chihō niokeru Sugi no Suitai to Sansei Kōkabutsu niyoru Kanōsei (Acid Deposition and Japanese Cedar Decline in Kanto Region, Japan). *Shinrin Ritchi (Japanese Journal of the Forest Environment)* 28(1): 11–17.

Takano, Bokuichiro, Hiroko Saitoh, and Etsu Takano. 1994. Geochemical Implications of Subaqueous Molten Sulfur at Yugama Crater Lake, Kusatsu-Shirane Volcano, Japan. *Geochemical Journal* 28: 199–216.

Takeuchi, Ushio. 1971. Kōsuichu no SO_4^{2-} nitsuite (On the Sulfate Ion Content in the Precipitation). *Tenki (Weather)** 18(8): 19–22.

Tamaki, Motonori. 1985. Wagakuni no Usui no Kagakuteki Seijō (The Chemical Composition of Acid Precipitation in Japan). *Kankyō Gijutsu (Environmental Conservation Engineering)* 14(2): 132–146.

———. 1993. Shimin Undō nado niyoru Sanseiu Nettowaaku de no Sokutei Seika (Results from Citizen-Based Acid Rain Monitoring Networks).* *Kankyō Gijutsu (Environmental Conservation Engineering)* 22(12): 689–727.

———. 1997. *Watashitachi Nihon no Sanseiu Kenkyū Bunken Risuto (List of Japanese Acid Rain Research Documents).** Osaka: Kankyō Gijutsu Kenkyū Kyōkai (Environmental Engineering Research Society of Japan).

Tamaki, Motonori, Takunori Katou, Kyoichi Sekiguchi, Moritsugu Kitamura, Keisuke Taguchi, Mayumi Oohara, Atsuko Mori, Shinji Wakamatsu, Kentaro Murano, Toshiichi Okita, Yoshio Yamanaka, and Hiroshi Hara. 1991. Nihon no Sanseiu no Kagaku (Acid Precipitation over Japan). *Nippon Kagaku Kaishi (Journal of the Chemical Society of Japan)* 5: 667–674.

Tamaki, Motonori, Takatoshi Hiraki, and Masahide Aikawa. 2000. Progress in Acid Deposition Monitoring Technology in Japan. *Global Environmental Research* 4(1): 25–38.

Taniyama, Tetsurō. 1989. *Osoreru-beki Sanseiu (The Dread of Acid Rain).** Tokyo: Godo Publishing Co.

Thomas, William L., ed. 1956. *Man's Role in Changing the Face of the Earth.* Chicago: University of Chicago Press.

Tsuru, Shigeto. 1999. *The Political Economy of the Environment: The Case of Japan.* Vancouver: UBC Press.

Tsuruta, Haruo. 1989. Higashi Ajia no Sanseiu (Acid Rain in East Asia).* *Kagaku (Science)** 59(5): 305–315.

Turner, B. L., ed. 1990. *The Earth as Transformed by Human Action: Global and Regional Changes in the Biosphere over the Past 300 Years.* Cambridge, UK: Cambridge University Press.

Uemura, Shinsaka. 1993. Shimin ni yoru Sanseiu Chōsa: Chisana Uekibachi kara Sekai no Kankyō Mondai ga Miete Kuru (Citizen's Acid Rain Survey: Seeing Environmental Problems from a Small Home Planter).* *Kankyō Gijutsu (Environmental Conservation Engineering)** 22(12): 698–703.

Ui, Jun. 1991a. *Kōgai Jishu Kōza 15 Nen (15 Years of Lectures on Pollution Problems).** Tokyo: Akishobō.

———. 1991b. *Yanaka-Mura kara Minamata, Sanrizuka e—Ekorojii no Genryū (From Yanaka Village to Minamata and Sanrizuka—the Rise of Ecology).** Tokyo: Shakai Hyōronsha.

———. 1992. *Industrial Pollution in Japan.* Tokyo: United Nations University Press.

Uno, Itsushi. 1995. Observational and Numerical Studies of Long-Range Sulfate Transport in East Asia. In *1st International Joint Seminar on the Regional Deposition Processes in the Atmosphere (20–24 November: Seoul, Korea)*, 173–183. Seoul: Atmospheric and Environmental Research Institute of Seoul National University.

———. 1997. East Asia Scale Long-Range Transport Model coupled with Regional Atmospheric Modeling System (RAMS): Application to Early Summer Rainy Season. In *3rd International Joint Seminar on the Regional Deposition Processes in the Atmosphere (5–7 November 1997: Nara, Japan)*, 120–130. Kyoto: Disaster Prevention Research Institute of Kyoto University.

Uno, Itsushi, Atsuko Mori, Akira Utsunomiya, and Shinji Wakamatsu. 1998a. Baiuki no Higashi Ajia Sukeeru no Chōkyoriusō Kaiseki (Numerical Analysis of Sulfate High Concentration Observed during the Baiu Season).* *Taiki Kankyō Gakkai Shi (Journal of the Japanese Society of the Atmospheric Environment)** 33(2): 109–116.

Uno, Itsushi, Kentaro Murano, and Shinji Wakamatsu. 1998b. Shunki no Idōsei Kōkiatsu Tsūkaji no Niji Taiki Osenbushitsu no Chōkyoriusō to Henshitsu Katei no Sūchi Kaiseki (Numerical Analysis of Secondary Pollutants Transportation/Transformation Processes during a Spring High Pressure System).* *Taiki Kankyō Gakkai Shi (Journal of the Japanese Society of the Atmospheric Environment)** 33(3): 164–178.

Uno, Itsushi, Toshimasa Ohara, and Kentaro Murano. 1998c. Simulated Acidic Aerosol Long-Range Transport and Deposition over East Asia: Role of Synoptic Scale Weather Systems. In *Air Pollution Modeling and Its Application XII*, eds. Sven-Erik Gryning and Nadine Chaumerliac, 185–193. New York: Plenum Press.

Uno, Itsushi, and Seiji Sugata. 1998. Nihoniki no Shunki no Tairyūken Ozon Kōnōdo no Shimyureeshon (A Simulation of Springtime Tropospheric Ozone Episode over Japan Area).* *Tenki (Weather)** 45(6): 425–439.

van der Sluijs, Jeroen, Josée van Eijndhoven, Simon Shackley, and Brian Wynne. 1998. Anchoring Devices in Science for Policy: The Case of Consensus around Climate Sensitivity. *Social Studies of Science* 28(2): 291–323.

van Eijndhoven, Josée, and Peter Groenewegen. 1991. The Construction of Expert Advice on Health Risks. *Social Studies of Science* 21(2): 257–278.

Venkatram, A., P. K. Karamchandani, and P. K. Misra. 1988. Testing a Comprehensive Acid Deposition Model. *Atmospheric Environment* 22(4): 737–747.

Wang, Wenxing. 1993. Study on the Factors of Acid Rain Formation in China. *China Environmental Science* 9(5): 367–374.

Wang, Wenxing, Shaoxian Hong, and Tao Wang. 1993. Anthropogenic Emission Inventories of SO_2 and NO_x in China. A paper presented at the International Workshop on the Harmonization of Monitoring Techniques of Acid Deposition and Methodology of Emission Inventories of SO_2 and NO_x in East Asia, 27–29 January, Tsukuba, Japan.

Wang, Wenxing, and Tao Wang. 1995. On the Origin and the Trend of Acid Precipitation in China. *Water, Air and Soil Pollution* 85: 2295–2300.

Watanabe, Masao. (1990 [first published in Japanese in 1976]). *The Japanese and Western Science*. Translated by Otto Theodor Benfey. Philadelphia: University of Pennsylvania Press.

———. 1997. *Science and Cultural Exchange in Modern History: Japan and the West*. Tokyo: Hokusen-Sha.

WCED (World Commission on Environment and Development). 1987. *Our Common Future*. New York: Oxford University Press.

Weller, Phil. 1980. *Acid Rain: The Silent Crisis*. Kitchener, Ontario: Between the Lines Press.

Wetstone, Gregory S., and Armin Rosencranz. 1983. *Acid Rain in Europe and North America: National Responses to an International Problem*. Washington, DC: Environmental Law Institute.

Whelpdale, D. M., A. Eliassen, J. N. Galloway, H. Dovland, and J. M. Miller. 1988. The Transatlantic Transport of Sulfur. *Tellus* 40B: 1–15.

Wilkening, Kenneth E. 1996. The Hitachi Copper Mine and Acidic Deposition Problems in Early Modern Japanese History. In *Proceedings of the International Symposium on Acidic Deposition and its Impacts, 10–12 December 1996, Tsukuba, Japan*, ed. Kenichi Satake, 249–256. Tsukuba, Japan: Center for Global Environmental Research, National Institute for Environmental Studies of the Environment Agency of Japan.

———. 1999. Culture and Japanese Citizen Influence on the Transboundary Air Pollution Issue in Northeast Asia. *Political Psychology* 20(4): 701–723.

———. 2001. Trans-Pacific Air Pollution: Scientific Evidence & Political Implications. *Journal of Water, Air and Soil Pollution* 130: 1825–1830.

———. 2004. Localizing Universal Science: Acid Rain Science and Policy in Europe, North America, and East Asia. In *Science and Politics in the International Environment*, eds. Neil Harrison and Gary Bryner. Lanham, MD: Rowman & Littlefield.

Wilkening, Kenneth E., Leonard A. Barrie, and Marilyn Engle. 2000. Trans-Pacific Air Pollution. *Science* 290: 65, 67.

Wong, Anny. 2001. *The Roots of Japan's International Environmental Policies*. New York: Garland.

Yagishita, Masaharu. 1995. Establishing an Acid Deposition Network in Asia. *Water, Air and Soil Pollution* 85: 273–278.

Yambe, Yoshito. 1973. Tokyo Tonai niokeru Jumoku Suitai no Jittai (Declining of Trees in Tokyo). *Ringyō Shikenjō Kenkyū Hōkoku (Bulletin of the Forestry and Forest Products Research Institute)* 257: 102–107.

———. 1978. Toshiiki niokeru Kankyō Akuka no Shihyō toshite no Jumoku Suitai to Biseibutsusō no Hendō (Declining of Trees and Microbial Florae as the Index of Pollution in Some Urban Areas). *Ringyō Shikenjō Kenkyū Hōkoku (Bulletin of the Forestry and Forest Products Research Institute)* 301: 119–129.

Yoshida, Katsumi. 1971. Sanseiu to Asagao (Acid Rain and Morning Glories).* *Taiki Osen Nyūsu (Air Pollution News Report)* 66(1): 1.

Yoshimura, Shinkichi. 1932. Bandai Kazan Shishū no Kazanko no Chikoshōgaku teki Yosatsu Kenkyū (Reconnaissance of Regional Limnology of the Lake surrounding Volcano Bandai, Hukushima—Part 3). *Chirigaku Hyōron (The Geographical Review of Japan)* 8(12): 933–976.

———. 1933. Kata-muna, a Very Strong Acid-Water Lake on Volcano Katanuma, Miyagi Prefecture, Japan. *Archiv für Hydrobiologie* 26: 197–202.

———. 1937. *Koshōgaku (Limnology).** Tokyo: Sanseidō.

———. 1942. Sekai Koshōgaku no Kinyō (The Present State of Limnological Research in the World).* *Kagaku (Science)** 12(10): 370–373.

Yoshitake, Takashi. 1978. Tarumae Chiiki no Sutorōbumatsu Zōrinchi ni Hasseishita Ijō Rakuyō no Keitai nitsuite (Unusual fall of needles of Pinus strobus planted in the area of Tarumae in Hokkaido). *Ringyō Shikenjō Hokkaidō Shijō Nempō (Annual Report of the Hokkaido Branch of the National Forestry Research Institute)*: 70–75.

Yoshitake, Takashi, and Hisao Masuda. 1980. Tarumae Chiku no Zōrinchi ni Hasseishita Ijō Rakuyō nitsuite (On the Unusual Defoliation in a Plantation in the Tarumae Region).* In *Nihon Ringakkai Hokkaidō Shibu Kōenshū—Dai 29 gō) (29th Annual Conference of the Hokkaidō Branch of the Japanese Society of Forestry),* Rishiri, Hokkaidō*, 113–115. Sapporo: Nihon Ringakkai Hokkaidō Shibu (Hokkaido Branch of the Japanese Society of Forestry).

———. 1986. Tomakomai Chiiki ni okeru Sutorōbumatsu nado no Ijō Rakuyō ni Kansuru Kōsatsu (Study on Unusual Defoliation of Pinus strobus etc. in a Region of Tomakomai). *Ringyō Shikenjō Kenkyū Hōkoku (Bulletin of the Forestry and Forest Products Research Institute)* 337: 1–28.

Zhao, Dianwu. 1994. Monitoring and Research of Acid Rain in China. A paper presented at the Workshop on the Acid Rain Network in South, East and Southeast Asia (ARNSESEA), 17–19 May, Kuala Lumpur, Malaysia.

Zhao, Dianwu, and Bozen Sun. 1986a. Air Pollution and Acid Rain in China. *Ambio* 15(1): 2–5.

———. 1986b. Atmospheric Pollution from Coal Combustion in China. *Journal of Air Pollution Control* 36(4): 371–374.

Zhao, Dianwu, and Jiling Xiong. 1988. Acidification in Southwestern China. In *Acidification in Tropical Countries. SCOPE Report 36*, eds. Henning Rodhe and Rafael Herrara, 317–346. Chichester, England: Wiley.

Zhao, Dianwu, Jiling Xiong, Yu Xu, and Walter H. Chan. 1988. Acid Rain in Southwestern China. *Atmospheric Environment* 22(2): 349–358.

Index

Acid deposition, 34–38. *See also* Acid Deposition Survey; Japan; Policy; Science
agricultural crops and (Japan), 177–178, 261–262
aquatic organisms and (Japan), 261
computer modeling (*see* Atmospheric computer modeling)
cultural properties and (Japan), 213–214, 262, 284n19
ecological impacts and (Japan), 259–262
in Europe, 3–4, 36, 42, 43, 47, 84–85, 141–144, 168–171, 188–190, 281n11
forests and (*see* Forests)
historical chronology (Japan), 54–59
historical scientific momentum on (*see* Historical scientific momentum [Japan])
hot spots, 36, 44, 197, 268n3
inland waters and (Japan), 108–109, 181–182, 260–261
monitoring (*see* Monitoring programs [Japan])
in North America, 3–4, 36, 42, 43, 47, 85–86, 141–144, 168–171, 190–192, 281n11
policy (*see* Acid deposition policy [Japan])
science (*see* Japan; Science)
soils and (Japan), 261
standard problem-framework (*see* Problem-framework)
Acid deposition policy (Japan)
Period 1 (smelter smoke), 88–92
Period 4 (moist air pollution), 167–168
Period 5 (ecological research), 185–188
Period 6 (East Asian transboundary air pollution), 197–202, 207–210, 211, 219–222 (*see also* Regime)
Acid Deposition Survey, 168, 180–182, 203, 212, 237, 238, 240, 259
establishment of, 173–174, 231
long-range transport of air pollutants, 205–206
Acid rain. *See* Acid deposition
Acid Rain Survey. *See* Acid Deposition Survey
Actor-network. *See* Latour, Bruno
Air pollution (Japan), 121–123, 197–198. *See also* Kobe; Nagoya; Nitrate; Nitrogen oxides; Osaka; Sulfate; Sulfur dioxide; Tokyo
laws (*see* Pollution laws [Japan])
monitoring network, 124, 133, 180
transboundary (*see* Transboundary air pollution)
Akimoto, Minoru, 97–98
Alkaline substances in atmosphere. *See* Yellow sands
Arahata, Kanson, 66, 270n3

Ashio copper mine, 62, 63–68, 88, 90, 93, 95, 234, 270n3
Atmospheric computer modeling. *See also* Central Research Institute of the Electric Power Industry (CRIEPI); RAINS
 early models in Japan, 178–179
 long-range transport models, 40, 46, 158, 178–179, 206, 252–255

Background Air Pollution Monitoring Network (BAPMoN). *See* Meteorological Agency of Japan (MA)
Baiu (plume rains), 46, 256–257, 269n9
Besshi copper mine, 62, 63, 68–71, 87, 272n6
Boundary objects, 25–26, 268n10
Bridging objects, 25–29, 225. *See also* Boundary objects
 classes of, 26, 28–29
 creation of, 54, 233
 definition of, 26, 31
 identification of, 27–28
 influence on politics/policy, 44, 230–231
 Period 1 (smelter smoke), 91–92
 Period 2 (urban precipitation chemistry), 119
 Period 3 (acidified atmosphere), 145
 Period 4 (moist air pollution), 166
 Period 5 (ecological research), 184–186
 Period 6 (East Asian transboundary air pollution), 207–209, 210, 211, 213, 218–219, 238, 249, 252, 258, 259, 263
 science-to-policy, 26–29, 31, 189, 229–232
 science-to-science (or universal), 20, 26–27, 31, 37, 145, 157, 166, 184
Brundtland, Gro Harlem, 12

Canada, 43, 85–85, 144, 171, 190, 191
Central Meteorological Observatory (Tokyo), 105
 Chemical Analysis Section of, 98, 116, 118, 133
Central Research Institute of the Electric Power Industry (CRIEPI), 202, 204. *See also* Fujita, Shinichi; Ministry of International Trade and Industry (MITI)
 Acidic Deposition Project, 204, 224, 237, 238, 240–241, 259
 agricultural crop experiments, 177–178
 atmospheric computer models, 206, 252–254
 emission inventories, 249–250
 Inland Sea project, 174
 Kyushu monitoring project, 179–180, 206
 monitoring networks, 213, 243–245
Chemical Analysis Section. *See* Central Meteorological Observatory (Tokyo)
China, 43, 45, 49–50, 198, 219, 269nn8–9
 acid deposition and, 46, 213, 214–216
 EANET and, 220, 222
 emissions, 210–211, 215, 231, 252
 environmental problems, 201
 long-range transport from, 211–212, 216, 217, 245, 253, 256, 257, 258
Citizens (Japan)
 input into policymaking, 200–201
 monitoring programs, 44, 188, 282n8
 movements, 124–125, 126, 132, 150
Compensation. *See also* Pollution laws (Japan)
 smelter smoke and, 69–70, 71–72, 88–89
Computer modeling. *See* Atmospheric computer modeling
Context (cultural, political, historical)
 acid deposition in Japan and, 48, 231, 208–209, 232

bridging objects and, 29, 229, 230, 231, 232
science in Japan and, 83
Copper and copper mines, 41, 62–76, 273n15. *See also* Ashio copper mine; Besshi copper mine; Hitachi copper mine
Copper Mine Poisons Investigation Committee(s), 67, 69, 70, 72, 88, 274n27
Council of Ministers for Global Environmental Conservation, 195, 207, 208, 238
CRIEPI. *See* Central Research Institute of the Electric Power Industry (CRIEPI)
Critical loads, 189–190, 224, 234
Culture. *See also* Japan
as context (*see* Context)
cultural properties, 213–214

Desulfurization technology, 67, 70, 88, 125, 126, 128, 132
Dry deposition, 34, 40, 135
early measurements of, 104–105, 111–112

East Asian Acid Deposition Monitoring Network (EANET), 193, 211, 231
establishment of, 158, 219–222
Edo period, 63, 64. *See also* Sustainability
science and, 273n19
sustainability and, 92–93
Tokugawa shogunate, 62, 63
Emissions. *See also* China
inventories (Japan), 249–252
RAINS-Asia project, 249
England. *See* United Kingdom
Environment Agency of Japan (EA), 53, 168, 199–200, 209. *See also* Acid Deposition Survey; East Asian Acid Deposition Monitoring Network (EANET)
Air Quality Bureau, 131, 173, 200
conflict with MITI, 200
establishment of, 131, 199
Global Environment Division, 200
Institute for Global Environmental Strategies (IGES), 200
National Institute for Environmental Studies (NIES), 131, 150, 200, 204, 256, 279n4
Environmental policy, 11–13, 22–25. *See also* Policy; Policymaking
complementarity between science and, 12, 232–233
science and, 22–24
Environmental science, 11–13, 22–24. *See also* Science
complementarity between policy and, 12, 232–233
in Japan, 116, 117, 150
Episteme. *See* Foucault, Michel
Epistemic community, 21, 267n7, 268n11
Epistemic shift, 20–21, 31, 55, 113, 141, 163, 182–183, 197, 227–229, 234–235
Europe. *See* Acid deposition; European Monitoring and Evaluation Programme (EMEP); Germany; Norway; Sweden; United Kingdom
European Monitoring and Evaluation Programme (EMEP), 157, 158, 170
Expert community, 21–22
acid deposition in Japan and, 53–54, 202–205, 229–230, 237
activism, 29, 31, 229–230
definition of, 21, 31
Period 1 (smelter smoke), 83, 87
Period 2 (urban precipitation chemistry), 117
Period 3 (acidified atmosphere), 145
Period 4 (moist air pollution), 163, 167
Period 5 (ecological research), 183
Period 6 (East Asian transboundary air pollution), 202–205, 207, 214, 224
vs. scientific community, 267n8

Focus-problem, 16, 225, 226, 231
Forestry and Forest Products Research Institute (FFPRI). *See* Ministry of Agriculture, Forestry and Fisheries of Japan (MAFF)
Forests. *See* Germany; Japan
Foucault, Michel, 14–15
Fujimura, Joan, 17–18, 232
Fujita, Shinichi, 174, 179–180. *See also* Central Research Institute of the Electric Power Industry (CRIEPI)
Furukawa, Ichibei, 63–64, 67, 76, 271n4. *See also* Ashio copper mine
Furukawa mining company. *See* Ashio copper mine

Germany, 84, 123, 143, 170
  Waldsterben (forest death), 170, 189
Global Environmental Protection Research Plan (GEPRP), 207, 208, 238–239
Global Environmental Research Fund (GERF), 237, 238, 239, 240

Hara, Hiroshi, 133–134, 167, 173
Historical scientific momentum (Japan), 6, 117, 147, 167, 183, 200, 220, 232
Hitachi copper mine, 62, 63, 71–76, 86, 89, 90, 92, 272n8
Hitachi Electric company, 76, 186–187

Industrial revolution, 34, 35, 84
Itsushi, Uno, 256–257

Japan. *See also* Acid deposition; Acid deposition policy (Japan); Air Pollution (Japan)
  acid deposition historical chronology, 54–59
  acidified environments, 42–43
  climate, 43–47
  copper and copper mines (*see* Copper and copper mines)
  culture of forests/wood, 47–48, 188, 231
  culture of lakes, 43
  culture of rain, 1, 6, 188, 231
  environmental history, 54–59, 269n11, 270n12
  forests, 47–48, 159–160, 174–177, 185, 262
  lakes, 108–109, 181–182, 260–261
  mountains, 41
  nature and culture in, 38–54
  plate tectonics, 2, 40–41, 47
  politics and, 198–199 (*see also* Acid deposition policy [Japan])
  science and, 48–54
  seas, 40, 115
  soil and vegetation, 47–48, 182
  volcanoes (*see* Volcanoes)

Kaburagi, Tokuji, 75, 82, 87, 89, 92, 118, 272n13
Kaneyasu, Naoki, 210, 245–246
Kato, Saburo, 173–174
Katō, Takeo, 99, 104–105, 113, 118
Keidanren, 122, 195
Kellner, Osker, 76–77, 79
Knowledge brokers, 90, 230
Kobe
  dry deposition measurements, 104–105
  early precipitation chemistry, 99–100, 102–103, 104, 133–134
  Meteorological and Oceanographic Observatory, 98, 104
Kōgai, 127
Kohno, Yoshishisa, 177
Komeiji, Tetsuhito, 160–161, 167, 173
Koreas, 45, 46. *See also* North Korea; South Korea
Kōsa. *See* Yellow sands
Kosaka copper mine, 62, 63
Kuhara, Fusanosuke, 71, 72, 73, 272n9. *See also* Hitachi copper mine
Kuhn, Thomas, 14, 21, 226
Kurashige, Eijirō, 98, 102–103, 113

Lakatos, Imre, 14. *See also* Research program
Latour, Bruno, 21
Likens, Gene, 144, 157, 205
Long-Range Transboundary Air Pollution Convention (LRTAP), 166, 170, 189
Long-range transport of air pollutants. *See* Transboundary air pollution
Lucky Dragon, 108, 275n9

Matsudaira, Yasuo, 98, 113, 115, 118
Matsui, Hideo, 110, 113
Matsuki village, 66, 67, 89
Meiji period, 270n1
 science and, 79–83, 228
Meiji Restoration, 7, 55, 61–62, 63, 78, 225, 270n1
Meteorological Agency of Japan (MA), 199, 158
 atmospheric computer model, 255–256
 BAPMoN stations, 246–247
Minister of Economy, Trade and Industry (METI). *See* Ministry of International Trade and Industry (MITI)
Ministry of Agriculture and Commerce, 64, 69, 70, 72, 73, 76, 77, 89, 90, 271n5, 274n27
Ministry of Agriculture, Forestry and Fisheries of Japan (MAFF), 199, 209, 259
 Forestry and Forest Products Research Institute (FFPRI), 140, 159, 160, 176, 204
Ministry of Environment of Japan. *See* Environment Agency of Japan (EA)
Ministry of Foreign Affairs of Japan (MoFA), 199
Ministry of Health and Welfare (MHW), 122–125, 131, 278n16
Ministry of International Trade and Industry (MITI), 53, 122–125, 128, 131, 199, 209, 239
 CRIEPI and, 204, 241 (*see also* Central Research Institute of the Electric Power Industry [CRIEPI])
 Green Aid Plan, 211, 219, 224, 231, 284n17
 New Earth 21 project, 195
MITI. *See* Ministry of International Trade and Industry (MITI)
Miyake, Yasuo, 103–104, 105–108, 109, 112, 113, 115, 116, 117, 118, 133, 137, 158
Moist air pollution, 46, 178, 180, 200, 203
 investigation of, 153–158
 policy, 167–168
 problem-framework, 165–167
 trigger event, 152–153
Mongolia, 44, 45, 217, 220, 222
Monitoring programs (Japan), 242–249
 aircraft and ship, 245–246
 air pollution (*see* Air pollution [Japan])
 citizen (*see* Citizens [Japan])
 CRIEPI (*see* Central Research Institute of the Electric Power Industry [CRIEPI])
 ground-based, 245
 National Acid Deposition Monitoring Network (NADMN), 180, 243, 245

Nagoya, 62, 111–112, 137, 138
National Acid Precipitation Assessment Program (NAPAP). *See* United States
National Institute for Environmental Studies (NIES). *See* Environment Agency of Japan (EA)
Nishi-ga-Hara, 77–78, 87, 133
Nissan, 76

Nitrates (NO_3), 58
early measurement in Japan, 76, 77, 78
Nitrogen oxides (NO_x), 35, 36, 58, 59, 151, 173, 176, 198
ambient air quality standard, 131, 277n11
vehicle emissions, 162, 277n10, 286n10
Nongovernmental organizations (NGOs), 150, 201, 230. *See also* Citizens (Japan)
North America. *See* Acid deposition; Canada; United States
North Korea, 216, 285n22
Norway, 85, 143, 154, 168, 170. *See also* Acid deposition; Scandinavia

Odén, Svante, 142–143, 144, 157, 168, 205
Okada, Takematsu, 98, 103
Okita, Toshiichi, 147, 167, 173, 174, 206
sulfate research, 134–136, 160–161
volcanic emissions measurement, 138–140
Organization for Economic Cooperation and Development (OECD), 130, 168, 196
Osaka, 62, 151
early air pollution, 95–96
early precipitation chemistry, 77, 97–98, 102–103
Ozone (O_3), 198, 247

Pacific War. *See* World War II
Paradigm, 15, 21. *See also* Kuhn, Thomas
Particulate matter (PM), 131, 135, 151
pH, 44, 50, 136, 269n9
of lakes, 100, 181–182
prewar measurements of, 85, 97, 99–100, 102, 104–105
in Tokyo, 137, 158–159, 160
Plate tectonics. *See* Japan
Policy. *See also* Acid deposition policy (Japan); Environmental policy
complementarity between science and, 12, 232–233
definition of, 22–23
Policymaking, 24
Japan and, 198–199 (*see also* Acid deposition policy [Japan])
Pollution Diet, 130, 131, 132
Pollution illness (Japan)
"big four" pollution cases, 126, 129, 149, 150, 270n12
itai-itai disease, 126, 129, 278n16
Minamata disease, 126, 129
Yokkaichi asthma, 126, 129
Pollution laws (Japan). *See also* Tokyo
Air Pollution Control Law (1968), 128, 130, 131, 151, 173, 277nn10–11
Basic Law for Pollution Control (1967), 126–127, 128, 132
Pollution Damage Compensation Law, 131, 132, 277n12
Smoke and Soot Regulation Law (1962), 123–124, 128
Popper, Karl, 266n4
Problem-framework, 15–18, 225–227. *See also* Bridging objects; Science
acid deposition, 183–185
definition of, 15–16, 31, 266n3
epistemic shift in (*see* Epistemic shift)
Period 1 (smelter smoke), 82, 86–87, 90, 113
Period 2 (urban precipitation chemistry), 113, 116–119
Period 3 (acidified atmosphere), 141, 145–147
Period 4 (moist air pollution), 163, 165–167
Period 5 (ecological research), 183–185

Period 6 (East Asian transboundary air pollution), 197, 202, 205, 209, 210, 213, 214, 218, 224, 237, 238, 242–263
vs. scientific framework, 267n8
standard, 20, 31
standard acid deposition, 36–37, 39, 143, 183
standard vs. localized, 31

RAINS. *See* Regional Acidification, Information, and Simulation model
RAINS-Asia Project, 249, 284n18
Regime, 24–25
definitions of, 24–25
formation of, 24
Japan's foreign policy and, 193, 198, 211, 222–224
Regional Acidification, Information, and Simulation model (RAINS), 178, 190, 192, 224, 234
Research program, 15, 21. *See also* Lakatos, Imre
Russia, 217, 220, 222
Russo-Japanese War, 65, 74

Satake, Kenichi, 269n6, 279n4
Scandinavia, 166, 214. *See also* Norway; Sweden
Science, 13–15, 50, 79, 83. *See also* Acid deposition science (Japan); Environmental science; Problem-framework; Scientific capacity building; Scientists
complementarity between policy and, 232–233
environmental policy and, 11–13, 22–24, 226–227
Japanese culture and, 50–54
Japan and history of, 48–50
influence on policy, 229–232
localization of, 18–20, 33, 37, 38, 76, 83, 116, 226–227
Meiji period and, 79–83
studies, 13, 21, 266n2
packages of, 13–15, 225
universal and local, 18–20, 116
Scientific capacity building, 233
in East Asia, 228–229
in Japan, 92, 117, 120, 147, 228
Scientists, 50. *See also* Expert community
bureaucrats and (Japan), 52–53, 230
Japanese culture and, 50–54
Meiji period and, 79–83
Sekiguchi, Koichi, 160, 175–176
Sino-Japanese War, 65, 69, 78
Smith, Robert Angus, 84, 86, 273n21, 274n26
Solution path, 29, 31, 93, 120, 148, 171, 224
standard acid deposition problem-framework and, 38
sustainability and, 30, 233–235
South Korea, 216, 219, 220, 222
Standard package. *See* Fujimura, Joan
Sudbury, 85–86
Sugawara, Ken, 101, 111–112, 113, 117
Sulfate (SO_4^{2-}), 58, 197, 135–136
prewar values (Japan), 77, 97, 104–105
in Tokyo, 134–135, 137–138
Sulfur dioxide (SO_2), 35, 36, 42, 176, 197, 286n10
ambient air quality standard, 128
copper mines and, 62–63, 66–67
early plant experiments and, 75
emissions trading, 191–192
Sumitomo. *See* Besshi copper mine
Sustainability, 6–8, 11–13, 232–235
acid deposition in Japan and, 119–120, 147–148, 171, 192, 221, 222–224
crises in Japan, 7–8, 265n5
definitions of, 11–12
ecological, 12–13, 234
Edo period and, 7, 92–93
Meiji period and, 7, 78
social, 12–13, 234
solution path and, 30

Sustainability (cont.)
World Commission on Environment and Development (WCED) and, 11–12, 266n1
Sustainable development. *See* Sustainability
Sustainable societies. *See* Sustainability
Sweden, 136, 141–142, 143, 154, 168, 170. *See also* Acid deposition; Odén, Svante; Scandinavia

Taiwan, 216–217, 219
Tamaki, Motonori, 162, 167, 173, 203
Tanaka, Akamaro, 101
Tanaka, Shōzō, 91, 270n3
Tokugawa period. *See* Edo period
Tokyo, 62, 63, 65, 129
air pollution laws, 121–122, 128
air pollution (postwar), 140, 151, 251
air pollution (prewar), 77, 96, 107, 110
Central Meteorological Observatory (*see* Central Meteorological Observatory [Tokyo])
moist air pollution, 152–153
precipitation chemistry (postwar), 134–135, 137–138, 158–159, 160–161
precipitation chemistry (prewar), 102–103, 105–107
Tokyo Imperial University. *See* Tokyo University
Tokyo University, 76, 79, 82, 83, 100, 129, 273n18
Trail smelter, 86
Transboundary air pollution, 36, 275n8. *See also* Long-Range Transboundary Air Pollution Convention (LRTAP)
discovery of in East Asia, 205–207, 245–246
in East Asia, 5–6, 45, 193, 194–195, 196–197, 215, 216, 217, 218, 219, 222–224, 249, 252–259
in Europe, 165–166, 168, 170
in North America, 165–166

Ui, Jun, 129
United Kingdom, 81, 84–85, 123, 143, 170. *See also* Smith, Robert Angus
acid deposition (prewar), 97–98, 99
air pollution (prewar), 96, 107, 115, 273n21
industrial revolution and, 35, 84 (*see also* Industrial revolution)
United Nations (UN), 170, 221
United States, 101, 108, 123, 142, 144, 171
Clean Air Act of 1970, 191, 277n10
Environmental Protection Agency (EPA), 130
National Acid Precipitation Assessment Program (NAPAP), 171, 181, 191, 239
National Environmental Policy Act (NEPA), 127, 130
Unno, Kiyoshi, 74

Volatile organic compounds (VOCs), 252
Volcanoes, 216
emissions from, 138–140, 250
Japan and, 2, 41–42, 47

World Commission on Environment and Development (WCED). *See* Sustainability
World War I, 82–83, 272n12
World War II, 61, 108–109, 110, 121, 141, 265n5

Yambe, Yoshito, 140
Yanaka village, 65, 86, 274n27
Yellow sands, 45–46, 100, 205, 215, 216, 217, 259, 269n8
Yokkaichi, 124–126, 137, 148, 162
Yoshimura, Shinkichi, 100–101, 113, 116, 117
Yoshitake, Takashi, 159–160